DEVIL IN THE STACK

Also by Andrew Smith

Moondust

Totally Wired

DEVIL IN THE STACK

A Code Odyssey

Andrew Smith

Atlantic Monthly Press
New York

Drawings throughout by Ron Jones.

FIRST EDITION

Published simultaneously in Canada

Printed in the United States of America

First Grove Atlantic hardcover edition: August 2024

Library of Congress Cataloging-in-Publication data is available for this title.

ISBN 978-0-8021-5884-0
eISBN 978-0-8021-5885-7

Atlantic Monthly Press
an imprint of Grove Atlantic
154 West 14th Street
New York, NY 10011

Distibuted by Publishers Group West

groveatlantic.com

24 25 26 10 9 8 7 6 5 4 3 2 1

For those who would move
slow and fix things

Earlier physicists are said to have found suddenly that they had too little mathematical understanding to cope with physics; and in almost the same way young people today can be said to be in a situation where ordinary common sense no longer suffices to meet the strange demands life makes. Everything has become so intricate that mastering it would require an exceptional intellect. Because skill at playing the game is no longer enough; the question that keeps coming up is: can this game be played at all now and what would be the right game to play?

—Ludwig Wittgenstein, 1937

I wish to God these calculations had been executed by steam.

—Charles Babbage

Contents

Prologue 0: If

I remember the moment code began to seem interesting to me. It was the tail end of 2013, and in the excitable tech quarters of New York, London and San Francisco, a cult was forming around an obscure "cryptocurrency" called Bitcoin. We know the story well by now. The system's pseudonymous creator, Satoshi Nakamoto, had appeared out of nowhere, dropped his ingenious plan for near-untraceable, decentralized money into the web, and then vanished, leaving only a handful of writings and 100,000 lines of computer code behind. Who would do such a thing? And why? Like a lot of mesmerized onlookers, I decided to investigate.

There didn't seem much to go on, until a chance encounter in a coffee line at a Bitcoin meetup in the East End of London opened me to something new. The man I met was a Finnish programmer. He told me that while Satoshi had taken pains to cover his tracks, there were clues in his code if you knew how to see them. There were also antecedents: the Bitcoin mechanism was a work of genius, but its creator built on the groundwork of others, some of whom he had contacted during development. My adviser pointed me to an Englishman named Adam Back, one of a loose group of cryptographer hacktivists who came to prominence in the 1990s as self-styled "cypherpunks." I set off on the cryptographer's trail.

The cypherpunk agenda, when it appeared toward the end of the eighties, was at once simple and complex. Humanity's impending lurch online would be an epochal gift to anyone with a political or economic incentive to surveil, the rebel cryptographers warned. We stood at a fork in the technological road, with the broadest path pointing to an Orwellian future of industrial scale intrusion and forfeiture of privacy, in which no facet of our lives was too intimate to be colonized by anyone with the right programming skills. To repel the bad actors massing to swarm cyberspace, citizens would need tools in the form of cryptographic software. Cypherpunks aimed to supply these tools.

Essential to online privacy would be a payment system that mimicked the anonymity of cash by making transactions hard to trace. Feverish effort went into designing such a system, but the task was daunting. Code derives its power from being digital, at root *numeric* and therefore exactly and infinitely reproduceable. How would you make digital money that could be transferred at will but not copied; whose electronic movements were registered on a ledger but without recourse to a corruptible central authority? By the end of the 1990s most Cypherpunks had abandoned the quest as Sisyphean, even if some bright ideas were floated along the way. And the travails were not wasted. Eleven years after Adam Back described a cryptographic spam-filtering algorithm called "hashcash" in 1997, it would star in Satoshi's dazzling system. Now I learned that Back had been contacted by the Bitcoin inventor—anonymously, he said—to arrange attribution for the prior work.

Back was living in Malta, so we spoke on the phone. He was friendly and, on the surface at least, open, claiming to be as puzzled as me by the mystery of Bitcoin's founder. Yet the longer we spent sifting the evidence, the more surprised I was to feel my interest pivot from Bitcoin's wraith-like founder to the cosmos of code "he" inhabited. Up to that moment I knew nothing about computer code save that it consisted of light-speed streams of binary numbers a microprocessor could interpret as instructions. How a human engaged this datastream was obscure to me; how a deluge of numbers became action in physical space was beyond my imagining. Yet here were details being proffered. And they were dumbfounding.

I heard that coders used a range of "languages" to communicate with the machine, and that there were thousands of these human-computer creoles, including a few dozen major ones, that each had its own culture, aesthetic and passionate claque of followers. By this account programming languages were not only communication tools, but also windows into the world, ways of seeing and being with definable and sometimes conflicting epistemological underpinnings. When a programmer aligned with a language, all this baggage came with it. And when they sent a program into the world, the baggage went there too. For these reasons there could be rivalry bordering on animus between communities—a tension coders half-jokingly referred to as "religious wars" on the grounds that no one involved was ever going to change their mind or attachment. If the names of these languages tended to suggest either roses or unconscionably strong cleaning products (Perl, Ruby, COBOL, PHP, Go, Fortran...), their whimsy emerged from an electrifying "open-source" creative model through which coders provided their skills to the community, usually unsung and uncredited and with all results shared, free and owned by all in a way one prominent business executive decried as "communism."

Satoshi chose a language called C++ for the writing of Bitcoin. This was because the "C" family of languages offered little by way of shortcuts or safeguards for the naive or unwary. A no frills approach made C++ harder to use than most alternatives but consequently faster and more efficient—important in a system its creator hoped would become ubiquitous. One programmer likened C to a shotgun, powerful if it didn't blow your foot off, while Back and others discussed the Bitcoin code like learned exegetes, citing evidence that C++ was not Satoshi's "native" language, or that he had learned to code in the 1980s, just as one might with a literary text. Programmers sprinkle comments throughout their code for the edification or amusement of peers, and Satoshi had been found to wander between US and British spellings, suggesting that "he" could be "they." By the end of my inquiry I tingled with questions about the coder's singular art.

Neither I nor anyone else figured out who "Satoshi Nakamoto" was at that time—at least not in public. But for me the story didn't end there. I'd written about Bitcoin and blockchain and imagined my brush with code done until, a couple of months later, someone reached out on Twitter. They knew who "Satoshi Nakamoto" was, they told me, and none of the individuals discussed so far fit the bill. On the contrary, "he" was a trio consisting of two Russian coders led by an Irish mastermind who'd studied computing at a Siberian technological university and now worked for Russian state media. My informant purported to be a Brazilian male model whose previous girlfriend dated one of the Russians. He didn't know much, he said, but could provide a first name for one Russian and full ID for the Irishman.

The latter was real. And active on social media. Wary but intrigued, I settled in to watch for clues. My Twitter source also existed offline, but when I asked to speak or meet, he stalled before following Satoshi back into the microcosmic ether. Aspects of his story made sense. One of Bitcoin's key promises to supporters—of the political left *and* right—was to upend the global financial order, an upheaval likely to serve Russia well. For multiple reasons Bitcoin was easier to credit to a team than an individual. And yet crucial details of what I heard were impossible to affirm or rebut. Needing perspective, I called an ex-colleague who now edited BBC TV's flagship current affairs program. Like me he was cautious, suspicion complicated by inability to see a motive for such an elaborate hoax, if that's what this was. We arranged for me to meet with Thomas Rid, then professor of Security Studies at Kings College London, whose research and writings had established him as an influential thinker on cybersecurity. Rid pledged to bring a Russia expert associated with the British intelligence agency, GCHQ.

We met in the windswept grounds of Somerset House on the north bank of the Thames, the kind of assignation point favored in the novels of John le Carré and Graham Greene, Cold War thrillers I'd devoured as a child and was disconcerted to find returning to currency. The experts arrived and I described what I was seeing. They listened, probed, tried to find precedents for this situation. None fit cleanly. Rid's companion saw Russian markers in the choice of a "Brazilian model" as conduit: someone not so exotic as to be absurd but distant enough to resist validation. This meeting was early in 2014, so little was understood about Vladislav Surkov, the Americanophile former theater student and advisor to Russian president Vladimir Putin, a fan of Black Sabbath and Tupac Shakur who quoted Ginsberg by heart while treating efforts to demolish American culture as an intellectual parlor game. To this end Surkov had developed an infowar technique called reflexive control, by which one entered the mind of a foe completely enough to feel their anxieties, fears, longings and delusions, learning to tweak these vulnerabilities until the target was ready to turn on itself. In this scheme, anything that undermined trust in institutions like the media was worth pursuing. Planting stories to be debunked was an anchor of Surkov's strategy.

But the big picture wasn't focused yet. We decided the Russian secret service was directly or indirectly behind the Twitter approach, aiming to hone the mystique of a leader who had made his country a poster child for kleptocratic dysfunction. So I reported back to the BBC and we let the matter drop. Only later did I see how my focus on Bitcoin blinded us to the story staring us in the face and primed to define the next decade. Just as the cypherpunks had predicted, computer code was seeping unchallenged and at an accelerating rate into every area of our existence. Within a few short years almost nothing any of us did would happen without it. And the world didn't seem to be getting better.

Prologue 1: Then

Over the next few years my attention was repeatedly drawn back to the demimonde I'd glimpsed through Bitcoin. The exoticism of code culture and its haunting alien logic remained on my mind, but more compelling were suggestions of a causal link between it and a fast-deranging human environment. The flood of code had intensified, and while most of the software embodying this code did things we liked, a growing proportion allowed online terrorists to spread viruses; businesses to cheat regulation; criminals to coin new forms of theft; data-hoarding digital monopolies to grow like mold and replace settled industries with precarious, loss-making ones, leaving civil society febrile and unnerved. The electoral chaos of 2016 showed how little we understood the quasi-occult technical powers now vested in a few hands—and how the work of these unseen hands could combine erratically at scale. By 2018 and a near perpetual slew of scandals, it was clear society had a problem with the software being written to remake it. From certain angles, life could appear to be getting worse in eerie proportion to the amount of code streaming into it.

What was the problem, though? One possibility was also the most obvious: it was the cohort writing the code, overwhelmingly white and Asian men, happier in the company of machines than fellow humans, if you believed popular culture and TV shows like *Silicon Valley*, *Mr. Robot* and *Russian Doll*. The main characters in these fictions mostly conformed to a category of code savant called the "10Xer," meaning someone with ten times the average productivity, who could make a computer do extraordinary things while struggling to connect with people—the implication being that these two traits were linked. Was the misfit nerd stereotype accurate? And how did it relate to the fact that while most professions had moved toward openness and diversity since the 1980s, code had charged determinedly in the opposite direction, reaching a nadir in 2015 when only 5 percent of programmers identified as women and fewer than 3 percent were Black . . . a low from which it has barely shifted? Coders were rebuilding our world for us without having posted a spec sheet: Could they be consciously or unconsciously recasting it in their own image, privileging their needs and assumptions above others? Would it be more surprising if they were, or weren't? And did any of this connect to a Silicon Valley ethos springloaded to recklessness? It began to seem important to ask who these destroyers and rebuilders of worlds were, what they thought they were building and why.

Or were coders incidental to our code problem, just functionaries following orders within a debauched business milieu? It was hard not to notice that when UN investigators charged Facebook with abetting genocide in Myanmar in 2018—in addition to stoking fatal violence in India and Sri Lanka—the company's share price remained buoyant. Division transpired to be remunerative. So, were killer code and its more subtly destructive variants predictable products of an irrational system? Would fixing the system do the same for its code? In this thought lay hope.

But between coders and the industry they served lay another possibility. I was beginning to hear murmurs that tech beasts like Facebook and Google no longer knew exactly how their algorithms worked or would play with others in the wild. More unnerving still, one of the clearest drifts in automating societies was toward polarity, antithesis as base state, in a way that invited comparison to the binary yes-no, true-false, zero-or-one code underneath. "Programmers like clearly defined boundaries between things," one practitioner explained to me, and while I didn't yet understand why this was, I was not alone in seeing a symmetric shift to rigidly defined boundaries between people and things in the social sphere—nor in considering it mystifying and destructive. The proliferation of software and bifurcation of society could be coincidence, I thought. But what if there was some hidden glitch in the way we compute that would inevitably discompose us? Which led to a pair of even more awkward questions. One, how would we know? And two, what would we, could we, do about it?

The software being written by a remote community of coders was reshaping society more dramatically than any technology since the steam engine. I was curious about how binary digits meshed with the world; how numbers became actions; whether this was the only way to compute. More urgent, however, was concern that ignorance of the digital domain's workings compromised my—and others'—ability to examine its expanding role in our lives as the lines between technology, politics, bureaucracy and personal space pixelated away to nothing. Over time I came to believe that the only way to communicate on equal terms with the mavens encoding the parameters of my life was to follow them Pied Piper-like into the microcosmos they were creating. And the only way I could see to follow them was by learning to do what they did. Or at least trying. Was a 10X-shaped "coding mind," conferred at birth, required to make sense of the microcosmos? I guessed I was about to find out.

In retrospect I should have had more qualms about entering the domain of code. But the qualm I did entertain was big.

Programming was notorious as a difficult endeavor that most novices walked away from. Assuming I could learn (an assumption I was in no way entitled to make), would computing's binary logic force *me* to become more binary? This was not a trivial concern. My life and work had been built on shades of gray and lateral motion of thought; on challenging "clearly defined boundaries" where I saw their semblance. Would immersion in code reprogram my cerebral operating system, narrowing its scope? In hopes of understanding any changes, I contacted a German team researching how the brain treats this weirdest new input, offering myself up to their study. To my delight the offer was accepted. If I was going to be digitized, I could at least do it consciously and in a way that shed light on the process.

CHAPTER

1

Revenge of the SpaghettiOs

I had nothing to offer anybody except my own confusion.

—Jack Kerouac, *On the Road*

The first six months I spend around code are the most disorienting of my life. If you want to communicate with a computer, you must learn one of the programming languages used to translate between them and us. A daunting challenge, you think. Until you encounter the hair-tearing torment of choosing a starter language in the first place, a task that by some trademark coder alchemy turns out to have no obviously right answer and yet hundreds of wrong ones. Credible guesses at the number of these human-machine argots, most written from some bewildering mix of altruism, curiosity and mischief, run from 1,700 to 9,000, with a precise figure no more calculable than species of bacterium in Earth's soil.

I spend weeks trawling the web for guidance on where to start; consult every programmer I know or have connection to—find all either hedging or contradicting what the last source said. Online threads seem to spin away forever without resolution, often exploding into rancor for reasons I can't discern, until I feel like a child watching my parents argue, wondering if what I'm hearing makes sense even to them. At first I hear just a litany of names linked to reputed attributes, demerits and specializations, none of which I am in a position to assess. With nothing to hold on to, I enroll in online bootcamps and courses more or less at random, hoping something will stick. But nothing does.

In this context, freeCodeCamp appears as if borne on a raft of light. Founded in 2014 by an idealistic schoolteacher-turned-programmer named Quincy Larson, fCC began with the explicit goal of broadening a shockingly shallow coder gene pool then hitching some of the newly diversified cohort to nonprofit work. To this end he established a free, user-friendly, step-by-step online course, complete with a vast and growing international network of self-organizing local groups and meetups. Larson's theory was that by rationalizing the learning process and establishing an entry-level "canon" of languages and software development tools to learn, guesswork could be reduced. My sense of gratitude would be hard to overstate.

The fCC curriculum will burgeon in coming years but for now revolves around a trio of languages at the heart of the web—HTML, CSS and JavaScript. Each member of this trinity works inside web browsers to render pages for human consumption. HTML (hypertext markup language) delivers content and structure, while CSS (cascading style sheet) describes how HTML content will look. The optional JavaScript is used to manage dynamics, or how any active features might behave.

HTML turns out to be a thrill. Written at the dawn of the web by Sir Tim Berners-Lee in 1990, its syntax is the digital dancing uncle at the wedding—awkward, but seldom scary. Its building blocks, like headers, subheads, paragraphs and images, are called *elements* and demarcated by opening and closing *tags* comprised of angle brackets and forward slashes. For example, a header is enclosed in the tags <h> for opening and </h> for closing. Header tags are also given a number from 1–6 to indicate their size, and anything appearing between them will be CSS code governing stylistic details of that element. Take the statement:

```
1 <h1 style="color:yellow; font-family:Helvetica; text-size=60px">
2     Do coders dream of numeric sheep?
3 </h1>
```

To a code-phobic eye this looks crazed, almost psychotic. But the words and symbols translate as "the element *Header 1* (h1) will be rendered in *yellow Helvetica, 60 pixels* high and read *Do coders dream of numeric sheep?*" Frightening only if you happen to be afflicted with aesthetic sense.

It becomes apparent that the collegial open-source development paradigm—through which anyone can contribute ideas and code to a project—has a consequence: the languages I am learning betray an almost comic absence of consistency or syntactic grace and a baffling panoply of ways to do things. I also get my first sense of how literal and pedantic computers are at the programming level, with a stray comma being enough to crash a system. One veteran programmer tells me she spent six months hunting a bug in a big program, knowing others had tried and failed, eventually tracing the problem to a dash that should have been an underscore. A *shift-devil*, she called it. This is why coders crave clearly defined boundaries and there is no such thing as "coding outside the box," she adds, because whatever occurs on the blind side of intention is a bug. To a coder, safety means straight lines and silos. And the exasperations don't end there. Sometimes languages become extinct and yet remain in the bowels of large programs as "braindead" or "zombie" code. Modern programmers rappel down, see the old code is there and still doing work even if they don't know how or why, then creep back to the surface before anything can break. No wonder coders can appear crotchety to civilians. My fast but hacky typing already looks like a vector for future pain.

But there is joy aplenty, too. Being able to manipulate a basic simulated web page in fCC's code editor is exciting. I soon learn that by simply right-clicking any web page I can inspect and even change its underlying HTML and CSS to alter the appearance of the page. There's naughty fun to be had going to Google's homepage and changing "Google Search" to "Google Schmearch," coloring it purple or rewriting the page in Cockney rhyming slang or jive. Most programmers describe a visceral rush the first time they were able to make a computer do their bidding: to my surprise I feel this too.[1]

Ability to change the appearance of a web page illustrates a deeper truth. It's easy for anyone raised on TV to assume all screen images are transmitted in the way of that medium. We now see that this is not how a web page works. When a user visits a site, their

---

1 Anyone with access to a computer can try this now: go to Google's home page, right-click it anywhere and select "Inspect" from the resulting pop-up menu: this will open an inspection bar at the bottom of the screen. From the menu at the top of the bar, click "Inspector" to see HTML code appear immediately underneath, while CSS will show in a sidebar to the far right. Now click the mouse icon immediately to the left of the "Inspect" button and hover your own mouse over any element on the Google page to see the code pertaining to it. Click on the element you want to mess with, go to the CSS and look for *keywords* like "color" or "font" or "height" and the *values* bound to them by a colon. Double-click the value (say, a blue color given as "#003eaa") and type an alternative of your choice (maybe type the word red or gold or aquamarine). The color of that element will change. Better yet, text intended to be "printed," meaning visible on the page, appears in the HTML, usually colored white and always skewered between the sharp ends of angle brackets (as per >Do coders dream of numeric sheep?< above). Double-click the text and rewrite it to your heart's content. The superlative Norwegian-run code education site W3Schools offers lists of HTML elements and CSS properties to watch out for and play with.

browser reaches out to the site's server and fetches nothing but raw code: there is no page as such. The browser then renders something that makes sense to a human, organizing all the page elements in real time according to instructions contained in the code, fetching specified fonts, images, video and data from wherever it might be on the web. The latter are held in place only for as long as they are being looked at, after which the "page" disintegrates. The real-world equivalent would be if every house on your street concealed a team of builders who raised it when you looked and took it down as you turned away. The question "What is a web page?" sits easily with "What is the sound of a tree falling in the forest if no one is there to hear it?" For the first time I understand why pages could take an hour to draw in the early years of the web. The instantaneity we expect now is what suddenly seems surreal.

Another revelation is that all those dot-coms whose names consist of two or more capitalized words run together, like DoubleClick and AirBnB, are not just trying to be cute: this style is called Pascal case and is commonly used in programming environments, where spaces ("whitespace" to a coder) are not always neutral. Make the first word lowercase (à la freeCodeCamp) and you have camel case. These styles can also be referred to as *UpperCamelCase* and *lowerCamelCase*. An identifier consisting of two or more words separated by an underscore (like_this) can be referred to as *snake case*, while substituting underscores with dashes renders *kebab case*. If a web developer doesn't yet know what written content will be used on part of a page, they use placeholder "lorem ipsum" text, drawn from a randomized portion of Cicero's first-century BC philosophical treatise *De finibus bonorum et malorum* ("On the Ends of Good and Evil")—which has been used by typesetters for this purpose since the sixteenth century. You quickly realize you've seen this "lorem ipsum" text before; it appears on a web page when its coder forgets to replace placeholder text with finished material.

To begin with, my learning goes well, with evening freeCodeCamp sessions forming a routine I look forward to as I start to have—whisper it—*fun*. The degree to which code colonizes my mind knocks me sideways: soon I wake up dreaming of CSS and thinking about it when doing other things; "seeing" solutions to problems the way a pianist might visualize melodies on a keyboard. I start to notice my wife Jan watching me with concern and am not sure if this is for me or her. All the same, confidence is high by the time I reach the curriculum's first solo project, to design and build a tribute page to someone I admire. "I think I'm going to be good at this," I hear myself declaim to the code gods one night.

Oh yeah? comes the reply. So far the freeCodeCamp course has involved learning basic syntax and ways to perform specific operations, like sizing and centering images, choosing and deploying rows of buttons, embedding links, managing fonts and so on. I thought I understood the fundamentals, but now I look at the blank code editor and something truly bizarre happens: my head empties in a way I have never experienced before. Stumped how to react I look again and—by some yet-to-be-explained quantum effect, I presume—my head empties still more. With the training wheels off I abruptly grasp that I know nothing and spend two whole sessions paralyzed, wondering where to start, feeling as adrift as I have ever felt in my life. So panicked am I by the sudden crash that I resort to radical measures and make my first good choice.

Down the phone from Texas, Quincy Larson of freeCodeCamp does his best to set me at ease, chuckling as he assures me my experience is typical. He forwards a funny graph entitled "Programming Confidence vs. Competence," which shows the two commodities moving in opposite directions at first, with an early peak in confidence followed by a precipitous fall and slow creep back as genuine competence is acquired.

"I think any able mind can do it," he tells me in response to my fear that I don't have the right kind of brain. "I view learning to code as primarily a motivational issue."

Then he delivers a truth I had never considered, one that explains code's power and difficulty; its uniqueness.

"The thing that gets lost, and which I think is important to know, is that programming is never easy," he says. "You're never doing the same thing twice, because code is infinitely reproducible, so if you've already solved a problem and you encounter it again, you just use your old solution. This means that by definition you're kind of always on this frontier where you're out of your depth. And one of the things you have to learn is to accept that feeling—of being constantly wrong and not knowing."

Which sounds like it could be a Buddhist precept. I'm thunderstruck.

"Well, constantly being wrong and out of your depth is not something people are used to accepting. But programmers have to," he concludes.

Relieved in the way of a rabbit who's outrun a fox to be flattened by a car, I follow Larson and freeCodeCamp's advice to join an extraordinary web community called Stack Overflow, the fortieth-most-visited site in the world at writing, where perplexed programmers canvas help from a global community of peers. A forbidding hierarchy of rules attaches to what kind of questions may be asked, and how—Lord save the blushing newbie who submits one that's been asked before—but most beginner issues have been encountered and dealt with already, folding into a giant database of solutions ranked for popularity and indexed in perpetuity, relieving the need to take pride in hand and frame something new. The question of why so many skilled coders donate time and energy to the community in this way, with no evident benefit to themselves as individuals, will grow to fascinate me. If coders are remaking the world, why doesn't the world look more like Stack Overflow?

At length, with copious help, the roadblocks fall away and I finish my tribute page to the late Apollo moonwalker Dr. Edgar Mitchell. My first web page is small, basic and likely to provoke unintended nostalgia for 1995 in anyone who was on the web then, but having been through the wringer I feel jazzed to have made it. Including wrong turns and becalmed interludes of despair, the learning process took several months. "So programming is hard but not impossible," I find myself telling people. Until one of those people drops a bombshell. Their programmer friend says HTML and CSS constitute coding but not programming. Trying to hide my irritation, I tell them they're wrong—they must have misunderstood what their friend said . . . only for them to come back insisting "No, he says programming is algorithms and there are no algorithms in HTML or CSS, you're just moving stuff around like fridge magnets—that's why you found them relatively easy."

My head swims, first because I didn't find any of this easy, then with cruel knowledge that even after months of toil I haven't understood programming well enough to understand how profoundly I don't understand it. Having always considered myself reasonably capable, code now exposes me as probably no smarter than an economist. It's humiliating. I probably focused on the languages I did *because* they're not algorithmic. Algorithms are where code and the new world encoded by it start to get wacky. There's no avoiding them any longer.

🐞 🐞 🐞

Algorithm: In the popular imagination this has come to mean a giant, almost supra-human entity that crunches lots of data to deliver things we like (news, music, friends), makes decisions about stuff we want (loans, education, jobs) or is out to get us like a monster from a 1950s B-movie. In reality all computer programs consist of algorithms, which are simply rules for treating data, defined by the word *if*. The simplest algorithm would consist of the statement "if *a* is true, then do *b*; if it's false, do *c*." If/then/else: the binary yes-no, true-false logic of computing, dictated by the eccentric workings of a microprocessor and projected outward from there. ***If** the customer orders size fifteen shoes, **then** display the message "Out of stock, Sasquatch"; **else** ask for a color preference.* How scary can this be?

I'm in luck, I think. At first. The third pillar of the web development trinity, JavaScript (JS), was written in the mid-1990s with the specific aim of bringing algorithms into a web browser. It has since become ubiquitous. But as I angle into freeCodeCamp's JS course my mind simply glances off it—a novel sensation for me; different from the one I had with HTML.

The first impediment to my progress is the profoundest. After a life in the fluid province of evolved biology, trying to "think" my way into the awkward sequential modus of the Machine—to *feel* its logic—is as uncomfortable as anything I've ever tried to do. The problem isn't algorithms as such. We use algorithms all the time in everyday life. ***If** Jan finishes work early enough, **then** we'll go see a movie; **else** we'll stay home and watch* Better Call Saul. But the computer expects every thought and action to be broken down into a sequence of individual instructions in a way that feels unnatural and *is* ahuman, this being a realm where there is no common sense or intuition to fall back on and *everything*, literally, is literal. What does this mean? If I say to a dinner guest, "Can you pass the salt, please?" they are probably moving before the question is finished, with multiple processes happening synchronously and no need for further elucidation. Had my guest been possessed by the spirit of a computer, however, my instruction would have sounded more like this:

"Uma: would you please access an image of a salt shaker from memory; scan the table for something similar; henceforth identify that object as "salt_shaker"; calculate whether salt_shaker is within reach; **if** no, **then** wonder why I asked you rather than Beyoncé to pass it; **else**, calculate whether it is nearer your left or right hand; assess whether there is any cutlery in that hand; **if** yes, **then** lay it on your plate; move your hand in the direction of salt_shaker; stop the hand over it; lower hand; grasp shaker; etc. . . ."

All while praying no bug impels my guest to hurl the shaker through a window or try to eat it. The further we travel up the stack toward higher-level languages like JavaScript, the more of this pedantry gets hidden. But not enough for my liking. I'm not sure I can think like this, or even want to, and any suggestion that I could learn to enjoy this process seems far-fetched right now. I feel like I'm trapped inside a game of *Tetris*.

The second assault on my equilibrium sounds petty but is real. Like many programming languages, JavaScript's syntax is influenced by the utilitarian C, with cluttered flights of brackets, braces and parentheses everywhere and each statement, however short, ending messily in a semicolon. They tell me JS is improving rapidly, but to my present way of seeing it looks like a child barfed a bowl of SpaghettiOs onto a screen. In my daily life I spend a lot of time thinking about syntax, trying to make it simple and clear, always in service to context. Put simply, I hate looking at JavaScript, which can't augur well for learning to write it.

And so I'm dumped back into what I've come to think of as the "coding mind" question: whether code is and always will be the natural habitat of a certain kind of mind, or even brain. Given the dramatic gender, race and class imbalances in the profession, allied to the power coders hold to reshape our lives to their own specification, this thought makes me nervous. It also implies that my rash dream of penetrating code culture was doomed from the start. Craving clarity either way, I approach someone who has thought deeply about the trials of programming. Gerald Weinberg's book *The Psychology of Computer Programming* was written as long ago as 1998 but is still considered the subject's foundational text. Via email I lay bare the disappointment I've felt; my fear that an elliptical, lateral-tending mind like mine simply can't do this. The speed and certainty of his response takes me aback.

"No, there are many ways to approach programming," he assures me by email. "The important thing for a programmer is to study your own mind and habits, then improve them as needed." He directs me to a blog post of a few days previously, in which he echoes Quincy Larson by elaborating:

> *A computer is like a mirror of your mind that brightly reflects all your poorest thinking. To become a better programmer, you have to look in that mirror with clear eyes and see what it's telling you about yourself. Armed with that information about yourself, you can then select the most useful external things to work on. Those things will be different for you than for anyone else, because your shortcomings and strengths will be unique to you, so advice from others will often miss the mark.*

Hence the lack of agreement over the best languages to learn. Hence Jun Wong of Hacker Dojo, the remarkable cooperatively run Silicon Valley startup incubator, confiding to me that while he can't predict whether a person will be a good programmer from speaking to them, he can venture what *kind* of coder they will or would be. My poorest thinking? Self-evidently I leap into things before thinking them through. Curious as to how other learners manage their discomfort, I find a slew of blogs and vlogs that combine to suggest coding is like watching *The Wire*: even future pros can experience multiple false starts. And this no longer surprises me. Over the months my sense has become less of adapting my brain to code's abnormal demands than of trying to build a new brain to run in parallel with the original . . . which sheds light on the problem but gets me no closer to finding a programming language that will solve it. At a loss what else to do, I revisit my early advisers and find the same patchwork advice as before. Until one day a coding colleague of a friend, a skilled C++ programmer and former musician, smiles sphinx-like on hearing my tale of woe. "I think I might know someone who can help you," he says.

CHAPTER 2

Holy Grail

Nicholas Tollervey had been one of those kids who wonder how their computer games work; whether it would be possible to customize them and make the machine do other cool stuff with a bit of code, the answer for him being *yes*. He was also a musician and took that path into adulthood, studying tuba at the Royal College of Music before noticing there were more astronauts in the UK than pro tubists—possibly more conjoined twins and pandas. So, he went back to his other early passions and now has degrees in music, philosophy, computing and education. I will learn that for reasons no one seems able to explain, the proportion of practicing musicians among coders is far higher than in the general population. Even so, Tollervey is no one's idea of the classic coder cowboy.

With me in the US and he in the UK, we meet on Google Hangouts. I learn that Tollervey has been a pro software developer for almost two decades; has worked on projects for companies like *The Guardian*, Freedom of the Press Foundation, NHS England, the Council of Europe and BBC; is fluent in a range of software tools and languages including C# and JavaScript. Most important to me, he is an emeritus fellow of the

Python Software Foundation (PSF), the US-based nonprofit that oversees development of the Python programming language. A few experienced developers have tried to steer me toward Python, so I explain my predicament and ask if the language might have anything different to offer me. He pauses while weighing his words.

"Well, I don't want to get into religious wars here," he chuckles, "so I should say that I enjoy JavaScript too. But yes, there is a reason people are suggesting Python."

He tells me the story of Python's emergence. How in the closing days of the 1980s, a Dutchman named Guido van Rossum decided to write a programming language that would be clear, concise and as easy to learn as possible, in which simplicity was paramount and transparency to other coders had the imperative force of a covenant. He named it Python after the British TV show *Monty Python's Flying Circus*.

Van Rossum was then working at the renowned Centrum Wiskunde & Informatica in Amsterdam, famous for having nurtured the legendary Edsger Dijkstra, software's cross between Albert Einstein and Jedi Master Yoda, and so a good base to work from. All the same, brilliant computing minds had dreamt of such a language for decades without success, even if the breadth and invention of their attempts turns out to be one of the great untold stories of the last eighty years, existing at the nexus of math, linguistics, philosophy, psychology, engineering, literature and neuroscience, as spectacular in its way as our first forays into space.

The lack of a definitive programming language turns out to be for a reason. Translating between the digital microcosmos and our analog human minds involves a fiendish tradeoff: make your language quicker and easier for the machine to process and it grows proportionally more alien to humans. Make it accessible and intuitive to us, however, and you commit the machine to lots of extra work—thereby handicapping the central processing unit (CPU) and introducing new layers of complication; of code to go wrong. Computerists use the metaphor of "The Stack" to describe this relationship. Lowest in the stack is machine code, the stuff that happens on a silicon chip, where tiny electrical switches called Logic Gates create and process the binary numbers we use to represent and ultimately manipulate the world. Just above that is Assembly Code, as used by early programmers at NASA and elsewhere, still present in the world but rarely engaged directly. At the very top are beginner languages like the child-friendly pictorial Scratch, where most of the machine's weirdness is hidden behind a colorful cartoon interface.

So it is that, in a typically precise inversion of everyday logic, coders employ the soubriquet "low-level" to describe the abstruse languages dwelling "close to the machine"—down by the silicon—and "high-level" to denote the more accessible ones further up the stack. JavaScript, Ruby and Perl are high-level, with lots of shortcuts and user-friendly guardrails designed to prevent serious errors. The more bracing C was considered high-level when it emerged from Bell Labs in the 1970s but became mid- to low-level as the stack expanded above it.

Python is a high-level *interpreted* language, meaning it consists of two parts: the human interface—the symbolic language we use to express our intentions, with its rules and conventions and syntax—and an *interpreter* that interprets our code for the machine. But its humane high-level visage does not make low-level code redundant: rather, the Python interpreter takes our work, which means nothing whatsoever to a microprocessor, and initiates a process of relaying it down the stack and into machine language. On the way down it will be converted, re-converted and processed several times—a process so labyrinthine and racked with exceptions, caveats and forks that a full accounting of it is hard to glean. Not all languages are interpreted: a lower-level language like C is compiled, meaning that our human source code is translated directly into executable machine code by a compiler, making it harder to use but efficient and therefore fast.

Computerists refer to this mind-bending arrangement as *abstraction*, an unassuming idea little understood outside programming, whose secret depths and full implications take many months to penetrate; even then will be replaced by nothing more comforting than a schizoid mix of wonder and fear. For now, I experience vertigo when trying to remind myself why this whole digital edifice doesn't just go *poof!* and vanish in a mist of flickering digits at any given instant.

I am not alone in being intimidated by the stack. On a trip to New York I drop in on the writer-coder Paul Ford, son of an experimental poet and author of a superb and improbably entertaining, novella-sized *Bloomberg* essay on code culture titled "What Is Code?" (sample: "Back in the 1980s, while the Fortran programmers were off optimizing nuclear weapon yields, Lisp programmers were trying to get a robot to pick up a teddy bear . . ."). In a conference room at his web design firm, Postlight, he grimaces when asked if even seasoned coders can hold the entire stack in their heads.

"That stuff is not comfortable, is it?" he offers in sympathy with my dislocation. "A mature programmer can go from very high to very low in the stack and explain how the pieces work. But that's maturity: there are really good software developers who, once you get below the level of what the web browser is doing, have no idea."

I already have an intimation that answers to the questions I'm asking are buried somewhere in the stack, mysterious as it is to me right now—and that, like a digital echo of Mr. Kurtz in Conrad's *Heart of Darkness*, I'm going to have to find my way down there, to where almost no one goes.

No one could have foreseen Python's future impact when Van Rossum set out to make it. As in showbiz—and to about the same degree—most new language entrants fall away and are forgotten. Written over a couple of years, the earliest Python was slow to catch on. Through the 1990s the coming high-level offerings were the chaotic JavaScript and prismatic Perl, and Van Rossum tells of being at a conference in San Francisco at the turn of the millennium, posting an invitation to a Python meetup and wondering if anyone would show, being relieved when 5 out of 12,000 attendees did. Nonetheless, a small community of believers formed and began to extend the language by building *libraries* of open source code prewritten to accomplish specific tasks, available to all users as importable *modules* or *packages*.

Only as the new century accelerated in tandem with the microprocessors driving it did the sagacity of the Dutchman's philosophic choices start to show. As software *became* the world, programs started to grow to a point where collaboration was unavoidable and transparency stopped being a luxury. Programmers would spend less time trying to decipher each other's code if the language being used was refined and unambiguous. If the lone wolf curmudgeon coder still exists, he is vestigial, Tollervey tells me—his days are numbered.

Had Van Rossum foreseen this shift? Or was he simply trying to make a language he wanted to use; build a community he would like to join? Either way, Python is not encouraging of showboaters: the primacy of communication, collaboration, community is etched into its DNA. Where JavaScript prides itself on a plurality of ways to approach a given problem (and is often likened to a coder Wild West by non-adherents), Python sets an ideal of there being one obvious way to do most things. "Pythonistas" see this attenuated freedom as a small price to pay for the sake of lucidity and accord. One prominent clan member will tell me that "To describe something as 'clever' is not considered a compliment in the Python culture." Making it easy to see how different personality types might be drawn to these very different languages, which appear to embody and express two distinct, and in many ways opposing, worldviews.

"Don't let anyone tell you code is not political," Tollervey concludes. "It contains all kinds of assumptions and ways of seeing, and the things we use it to make reflect these. You'll sometimes hear inexperienced developers boast about how many lines of code they've written. Well, a more experienced professional will brag about how many lines of code they've removed. An expert is someone who's able to say, 'This code doesn't need to be written—here's a solution where we don't need it.'"

Pythonistas admit that while their language isn't best at much, it's second best at almost everything. Improvements are floated via Python Enhancement Proposals, or *PEPs*, which anyone can suggest for the community to debate and action or not according to consensus. For most of the language's life Van Rossum, as officially annointed Benevolent Dictator for Life (BDFL), broke any deadlocks with a casting vote. A seminal PEP, PEP8, the *Python Style Guide*, was written by him in 2001. Another, PEP20, *The Zen of Python*, was contributed by an early enthusiast named Tim Peters and consists of nineteen declarations of principle. *Zen* is built into the language itself, ready to be called like a priest in dark moments. It begins:

Beautiful is better than ugly.
Explicit is better than implicit.
Simple is better than complex.
Complex is better than complicated.

Further down is a playful nod to Van Rossum:

There should be one—and preferably only one—obvious way to do it.
Although that way may not be obvious at first unless you're Dutch.

And my own favorite:

Now is better than never.
*Although never is often better than *right* now.*

A sentiment starkly at odds with the "move fast and break things" dogma coders at dysfunctional Big Tech firms have used to blow up so many functional predecessors, most especially those inclined to move slowly and fix things.

Tollervey laughs at my surprise in finding these thoughts embedded in a programming language, but by the end of our conversation I am sold. We discuss potential learning projects and I mention the idea of building a mini app to scrape Twitter for mentions of an author's books—my variation on a common beginner's project. He agrees this is doable and offers help if I need it, and for the first time in weeks I feel a glimmer of optimism. He also floors me with a suggestion: if I really want to see code in action and get a sense of who coders are and what they do, I should consider enrolling in the annual PyCon jamboree, which will host 4,000 Pythonistas, including him, in Cleveland. And before I can sweat the ramifications of such a tumble into the unknown, I hear a voice, belatedly identifiable as mine, agreeing.

CHAPTER

3

PyLadies and Code Freaks

Five a.m. The rail hub at Hopkins International Airport. Tired. Staring. How is it that while trains work the same everywhere, rail operators' ticket machines all find fresh ways to mess with your mind? A young man next to me stands frowning at the card the machine just spat back for the third time. It turns out he's here for his first PyCon, too. United by tiredness and irritation, we catch the train into Cleveland together.

Alex is from Seattle, a rank-and-file coder with one of the field's several classic origin stories. When a post-high-school life in fast food proved all it wasn't cracked up to be, he decided to try programming, and for several years drifted through languages including Ruby, JavaScript, Java and PHP without gaining much purchase, until he doubted his coding Right Stuff. Then he tried Python and liked it. These days he tends websites with the Python web framework Django.

As our train rattles through the outskirts of Cleveland toward a soft buttercup sun, this sounds like a story with a happy ending. But no, Alex says. The company he works for develops "AI" for use in fracking, an activity he doesn't approve of, and alternative jobs at his level are hard to land in tech-heavy Seattle. Disillusioned, he is thinking of giving up, with PyCon a last roll of the job-seeking dice. We pull into Tower City Station and exchange numbers, with a promise to catch up at some point during the conference. I roll away hoping he doesn't give up, because at the very least, he cares.

* * *

My tiredness reflects a significant life change. Cleveland turns out to be one of the more awkward journeys to make from the San Francisco Bay Area, to which Jan and I moved in the months prior to PyCon. For the next few years I will be where a large portion of the world's code is written, including much that is most contentious. With the cities of San Francisco and Oakland too pricey to contemplate we crossed the Golden Gate into the relatively rural county of Marin, to find an offbeat perch in the Bay Area's last remaining hippie enclave, a place where war stories from Woodstock and Monterey Pop abound and the only recognized crimes appear to be harshing mellows without a license and gluten smuggling. I quickly rue the way all music made after 1974 seems to ionize at the county border. And that, given the proximity of Oakland, the population is anomalously white, with half the townsfolk desperate to see this change and the other (often older) half to varying degrees not: twenty-first century America in tie-dye microcosm, and not uninteresting for that.

Another plus to Marin is being at the opposite end of the majestic San Francisco Bay from Silicon Valley, meaning there is almost no tech penetration—hence the lower prices—so when I need respite from functions and variables I only need step outside to hug a redwood or join an ecstatic dance group. At least that's what I think until I meet my immediate neighbor, Sagar, a programming sensei of around my own age at a major Valley firm, who on initial meeting seems to fit the classic nerdy coder stereotype to an almost comic degree, but soon proves lightyears from it—a dynamic I will grow accustomed to and never stop being amused by.

❦ ❦ ❦

The word *windswept* could have been coined for downtown Cleveland. As in other late capitals of mechanical industry, city burghers spared no expense in celebrating themselves with wide roads and long blocks and decorative open squares; with a legacy of grand civic buildings no one seems to enter or exit—relics of a time when power was projected outward rather than concealed in streams of numbers. For someone with no personal connection to Ohio I've spent a spooky amount of time in it, yet this is my first visit to post-industrialization's complicated Rust Belt exemplar. It's true what they say about Midwesterners being open and friendly by default, and I like the place instantly, even while wondering if Silicon Valley will be regarded similarly one day, as history implies it must, and what such a turn might look like when it comes.

Any such moment of reckoning feels infinitely remote inside the vast, subterranean Huntington Convention Center, which hosts Thursday night's PyCon reception party. If I'd expected the beard-stroking formality of most professional conferences, what I find more resembles a first day back at Hogwarts. I text Nicholas Tollervey and move to the cavernous main exhibition hall, where Pythonistas hug and exchange warm greetings while they drink and eat and buzz excitedly around sponsors' stands. To the fore I see the capacious berths of tech behemoths like Google and Facebook; behind me are a hive of startups whose tech-generic two-syllable names—*Hosho!*, *Zulip!*, *Tivix!*, *Nexmo!*—perennially suggest the work of either randomness or a committee of overtired toddlers. All, large and small, declare love for Python and an eagerness to recruit. At first I am flabbergasted to find recruitment pitches disproportionately directed at me, before realizing my age implies experience and/or influence. By the time I decide enjoying this unearned status may be a soupçon ethically louche, it's too late: I swagger off to meet Tollervey as the proud if confused owner of three job offers. Oh to be a programmer in the twenty-first century.

Nicholas and I greet like old friends. It takes me a while to understand why coding with other people creates such a bond of intimacy, but for now it's enough to marvel that so ostensibly analytic and rule-based an activity does. We chat easily as a stream of friends, acquaintances, strangers pirouette into our group before wheeling away somewhere else, as though everything is part of a giant program we're writing together. In a quiet moment I look around, noticing that while you wouldn't mistake the population of this giant room for a conference of lumberjacks or college quarterbacks, it could be any other collection of enthusiasts. A revelation: the international crowd here isn't just pro coders, they are also scientists and educators, public policy wonks, artists and everything in between. As expected, the crowd skews white or Asian and male, if much less heavily than the programming profession at large, and I will notice that speakers over the coming days more closely reflect the population outside.

"There's Guido."

Nicholas snaps me out of my reverie.

"The panjandrum. God, basically."

I follow his gaze to a man with a neatly trimmed beard, sporting black plastic-framed glasses not unlike the old NHS models once worn by The Smiths' singer Morrissey. He wears a neutral tweedy jacket and jeans, an amused smile softening already approachable features. In this company you never hear Guido van

Rossum's last name: he is "Guido," like Jay-Z or Rihanna, and if he had intended to sneak a bite in a quiet moment, as appears to be the case, a group of starstruck high school students have dashed that hope as they chatter and exclaim and take selfies with him. Over the three days of the conference I never see Van Rossum without a crowd of people seeking his attention, often just wanting to tell him how much Python means to them. Only occasionally do I see a fleeting sign of strain cross his face, quickly replaced by an encouraging smile.

Afterward, in good cheer, we head for a sports bar with a group including Naomi Ceder, soon to be retiring Chairperson of the Python Software Foundation (PSF), a remarkable human being who hammers a fresh nail in the coffin of my preconceptions. As a child in a Nebraska prairie town at a time when computers were seldom found in homes, she was fascinated by physics and astronomy but also by the ancient history in which she took her first degree, followed by a doctorate in Greek and Latin literature. Her entrée to computing mirrors that of Quincy Larson from freeCodeCamp: exposure to an Apple IIe computer while teaching in Greece ensured that when a staff computer at her later US school needed to be programmed, she was the natural candidate to learn.

When I ask Ceder how Guido foresaw programming shifts that ultimately attracted over ten million users to Python, she shoots back:

"He didn't- no one could! This may sound weird coming from a coder, but I think what Guido brings is an aesthetic sense. Not that he's not a gifted coder, because he is, but there are lots of those, it's not in itself a very distinctive quality. But his strong attention to the aesthetics of the language gave it a form and structure amenable to adaptation and scaling. Like a classical building."

Fascinating to me is the extent to which the community Van Rossum built appears to reflect values inherent in the structure of the language. And how, through an occult alchemy that often binds stable groups, these values are at once vague and understood by all. Ceder smiles at this observation.

"Well, yes. If you look at *The Zen of Python*, it's the nearest thing I have to a religious text, but it ends up being a feedback loop. People expect Python to be 'Pythonic.' But what is 'Pythonic?' It's like Python!"

I haven't heard the word "Pythonic" before.

"Oh, you will. The worst thing you can say to someone around here is 'That's not Pythonic.' Whether in terms of code or behavior."

"Can anyone become Pythonic?" I ask as our group breaks up in readiness for conference in the morning. "Could I?"

Ceder smiles. "Well, you know, people always used to say, 'Oh you're so good at this, how did you get so good?' And they never wanted to hear the answer, which was that I put in a lot of time puzzling over these problems. I always come back to this: don't try to learn everything. Learn the stuff that's useful and interesting to you and then branch out from there."

I will meet devotees of other languages who reject Pythonic assumptions with a passion—sometimes on *principled* rather than on pragmatic grounds. But for now I feel unexpectedly comfortable here. Soon I'm on the short walk to my digs, trying to assimilate the fact that I've had fun.

With between ten and twenty talks and events I could be attending at any given moment, the three days of PyCon pass in a blur of novelty and FOMO. Favorite technical sessions include

"Birding with Python" and "Love Your Bugs," in which a charismatic bug specialist named Alison Kaptur likens bug-hunting to a murder mystery (*Murder on the Outlook Express*, one presumes). Less orthodox are an inspiring session hosted by a collaborative, self-directed Brooklyn-based boot camp called The Recurse Center, where the phrase "Well, actually" (as in "Well, actually, that's not how you do it . . .") is *actually* banned, and "Coding through Adversity," where the programmer Chris Schuhmacher brings down the house with his story of learning to code while in San Quentin prison on a drug-related murder conviction, thanks to a pioneering program called The Last Mile. Further joys include an account of a family going off the grid aboard a converted schoolbus, Merry Prankster-style, with systems automated using Raspberry Pi mini-computers and showers contributed by Pythonistas across the country, and a woman named Q (for Qumisha Goss), who with wit and humanity describes the challenge of teaching Python to kids at a library in a poor district of Detroit. When Silicon Valley flooded Detroit with scooter and bikeshare fleets, she says, entrepreneurial kids figured out how to install lawnmower engines in them, leading to the Freakonomically impeccable phenomenon of bikeshares driving a spike in lawnmower theft.

"So, these kids are innovators," Q concludes, adding that the ability to program feels like a superpower to most of them. "And you can use your superpowers for good or evil. So, we let them test their superpowers, then we encourage them to do good. These kids have greatness in them, they just need the opportunity to develop it."

As I wander through the conference, by turns dazed and delighted, the ability to program seems like a superpower to me, too—perhaps more than it did when I set out on this journey. In the months since then I've been to meetups and auditioned countless online courses and tutorial books, at length settling on a fabulously engaging tome called *Python Crash Course*, by Eric Matthes. Now Matthes's book goes with me everywhere and the building blocks of programming are coming together. Key elements I've encountered so far include *collections*, *for loops*, *conditional* (or *if*) *statements* and *while loops*. The internal binary motion of the machine—*if-then-else, if-then-else, if-then-else, if-then-else*—still seems so foreign that these foundational concepts take time to internalize, and I feel like someone running for a bus that accelerates away whenever I get close. Only slowly do the fundamentals take shape in my mind, through use, to the point where I can explain them.

The most common collections are lists and dictionaries, which are exactly what they sound like. Pro coders have a colorful inventory of slang terms for situations they encounter: someone who obsesses over trivial detail is "yak shaving" (taken from the cartoon *Ren & Stimpy*) or "bikeshedding" (as in "Who cares if it's red or green—it's a friggin' bikeshed"), while getting hung up on problems extraneous to the main one is to be an "architecture astronaut." In Python, lists are denoted by square brackets, so a simple list of code slang could appear like so:

```
1 code_slang = [
2       "bug",
3       "hindenbug",
4       "heisenbug",
5       "nerdjack",
6       "pebkac",
7       "non-trivial",
8       "fubar",
9 ]
```

This snippet introduces another idea vital to code: the *variable*, in this case given the name "code_slang" and often likened to a named container in which a particular *value* may be stored (coders quibble about this metaphor, but it serves us for now). To assign values to variables, we use what most of us recognize from math class as an "equals" sign (=), but which in programming becomes the *assignment operator*. What the above code says is "take the given list of code slang terms and assign it to the variable called 'code_slang.'" From now on, wherever "code_slang" appears, it represents, or *symbolizes*, the list we have assigned to it, much as the word "tree" symbolizes a thing in the world without being that thing. The power of variables lies in the fact that we can change what they symbolize at will, using a range of tools built into the language. We might add or remove an item from our code_slang list, or put it into alphabetical order: the options are near endless.

Slang terms are of limited use without definitions, however, which is where dictionaries come in. Denoted by curly brackets, dictionaries hold items called *key-value pairs*. These could consist of states and their capital cities, soccer players and their clubs, or slang terms and their definitions, like so:

```
1  code_slang_dict = {
2      "bug": "a problem in a piece of software",
3      "hindenbug": "a catastrophic problem in a piece of
       software",
4      "heisenbug": "a bug that disappears or changes behavior
       when studied",
5      "nerdjack": "to hijack a conversation with excessive
       nerdy detail",
6      "pebkac": "problem exists between keyboard and chair",
```

```
7       "non-trivial": "a fiendishly difficult problem",
8       "fubar": "fucked up beyond all recognition",
9 }
```

The "key," which is really an identifier, is attached to the value it represents by a colon, with key-value pairs separated by commas. Once we understand that the value associated with a key can itself be a dictionary or a list, it's easy to see how illimitable these collections, which belong to a category called *data structures*, can be. We can now *iterate* over all or selected items in the collection, doing a prescribed piece of work on each—perhaps doubling numbers; making words lowercase; choosing specified types of items to work on and then send to a new dictionary. For this we would use a "for" loop, the "for" meaning "for each item in the specified collection, do *x*." Not all languages iterate this way, but Python does.

If statements and *while loops* are quintessential algorithmic devices. The former are also called *conditionals* because they define a condition or conditions under which an action should occur. We've seen this before: if *x* applies, then do *y*; else don't. If the price of a stock falls below five dollars, buy one hundred shares; otherwise leave it alone. By contrast, a while loop says, "While *x* condition applies, do *y*—and keep doing it until *x* stops applying." While the price of a stock remains under five dollars, buy ten shares, and keep buying ten-share increments until its price reaches five dollars, then quit the loop. The latter requires caution because if a condition for breaking a while loop is not defined with forethought, an *infinite loop* can occur and wreak havoc.

Where these code *objects* and processes come together is in the *function*, a block of code containing a reusable algorithm that automates a chosen task, the essence of programming. To give a simple example of what these algorithms look like, a weather app might benefit from a Fahrenheit-to-Celsius converter. Examples of this basic algorithm are all over the internet, based on the standard formula of taking a Fahrenheit temperature, subtracting 32 and then multiplying by 5/9 to arrive at a value for Celsius. A basic function to carry out this work in Python could look like this:

```
1 def to_celsius(fahrenheit):
2     celsius = (fahrenheit - 32) * (5 / 9)
3     print(celsius)
```

There are more concise ways to write a temperature converter in Python, but the more compressed the code becomes, the more cryptic it appears. What's going on here? I learn that in programming this code snippet is called a *function definition*, signaled by the keyword "def." Line 1 establishes the name we're giving to our converter function: "to_celsius." The parentheses following the function's name provide a portal for the input it needs to work on—the temperature we want converted—with the word "Fahrenheit" acting as a placeholder for the specific number we will provide whenever the function is *called*. In cases where a function doesn't require data to work with, the parentheses are left empty.

In line 2 we carry out the mathematical calculation and load the result into the variable "Celsius," while line 3 instructs the interpreter to print out the value stored in "Celsius" to the user. "Print" is one of a panoply of *built-in* functions contained within the language and ready to be *called* at any time. To call the function and provide a Fahrenheit temperature of 77 to convert, we would simply type:

```
to_celsius(77)
```

and expect to see an output of:

```
25
```

Magic!

The reason we compute like this turns out to be fascinating—and *involved*. Even at this stage I've seen enough to know that loops are elementary to the notion of programs, amplifying and concentrating actions in a way that allows a few lines to have vast impact. *Amplifying and concentrating*: Isn't that also a reasonable description of what the algorithms of Facebook, Google and Amazon; high-frequency traders on the stock market; rogue states and scrupleless businesses have been doing at the stygian margins of the human macrocosm? Food for thought going forward.

🐞 🐞 🐞

My most revealing glimpse into the code culture at PyCon doesn't involve code.

With an eye to imbalances in the trade, an ever-present conference theme is inclusivity. A clear Code of Conduct aims to make everyone welcome at conference, while a Spanish language track appears for the first time this year. PyCon profits underwrite regional conferences and meetups around the world, prodigiously so in Africa, and an active trans community has a voice extending to the very top of the organization, spearheaded by PSF Chair Naomi Ceder. But it comes as no surprise that a top priority should be encouraging more women into a field given as 93 percent male at writing. Central to this effort is a high-profile and well-supported group called PyLadies, whose annual fundraising auction takes place tonight at the convention center Hilton. I've wondered what kicking back would look like here, and this is my chance to find out.

With no idea what to expect I arrive early to find the ballroom filling fast. Seeing no one of my acquaintance, I head for a near-empty table at the far end of the room, where I meet Barry, a DC web developer and PyCon newbie like me, whose first career as an orchestral percussionist was swapped for linguistics and thence code. He rolls his eyes as he tells me his current job skews to JavaScript, which is evolving so fast right now that he's forced to learn new frameworks and tools every couple of months, never knowing if they'll still be around in two years' time. This can be exciting, but it's also exhausting, he says, as I wonder whether humans have ever before lived in such perpetual flux as coders do, and just how far down that road the rest of us will be happy to follow them.

Three further arrivals to our table work for Amazon in Seattle and are here to keep abreast of developments in this year's buzz application, Machine Learning, which Python is shaping to dominate. One of this triumvirate originates from Prague and is new to the company, but the eldest has been there long enough to remember the time of sixteen-hour days and such rapid expansion on the coattails of the internet that workers stacked desks into mezzanines until the Seattle fire department found out. He ditched a doctoral thesis in political science for programming, he says, which prompts me to note the high number of Pythonistas who've come from other careers, also reprising an earlier discussion about the disproportionate number of musicians in the fold, which I've never seen explained.

"I think that's because they both deal in abstraction," he muses, raising that word again. "I mean, I teach Python now and one thing I've noticed is that people whose minds deal easily with abstraction tend to take to it well. People with very visual minds often find it harder."

"Oh, like me!" interjects the Czech.

"Well, yes, but that obviously doesn't mean you can't do it, just that it can be more of a challenge."

We fetch drinks and join the long line for sliders and salad before it gets longer, returning to our table to find it a little fuller and a lot quirkier. Along with the quartet I already know, a still empty seat is flanked by two new men who could be any age at all, who stare at their food they eat in silence, occasionally gazing toward the stage in anticipation. Presently a third man I take to be their companion glides into the empty seat, carrying a plate on which four rigidly symmetrical, bunless sliders are arranged under architectural sheets of ketchup, mayo and mustard, topped by a canopy of fastidiously leveled cheese. For the remainder of an otherwise revelrous evening, none of this subgroup utter a word, betray an emotion or provide any evidence they know the rest of us are present at all: were they to blink and find themselves on a beach in Cancún or in space being chased by Klingons, there is no reason to believe their responses would be different. What to do? I follow the example of the Pythonauts around me, who remain unfazed. "You'll get used to it," Nicholas laughs when I recount the scene later.

And I do. Lifting eyes to the room, I take in the full breadth of humanity on view, expressed in an unanticipated multiplicity of subcultures. There are the geezers in Hawaiian shirts (in icy Cleveland!) and young tech pros in corporate hoodies or t-shirts, maybe a promo baseball cap; the company guys in indigo jeans and jackets. Many PyLadies rock ironic—I think—1950s frocks, against a sub-tribe of tattooed, shock-haired Euro cyberpunks who look like a different species in here, and hipsters who could be lighting techs for Arcade Fire or at least Maroon 5. In this company there is always at least one portly code vet in Victorian explorer garb, while beards run the gamut from Brooklyn dandy to downed Pacific theater pilot who doesn't know the war ended thirty years ago. Never far from view is a group I come to think of as the Code Freaks, meaning "freak" in the Frank Zappa and the Mothers rather than Mark Zuckerberg sense; typically older and relatively unisex, often vets of the first public-facing, counterculture-inspired computing revolution of the 1960s, '70s, '80s, who dress with a wizardly flamboyance exemplified by (but not limited to) long leather coats and big black hats and t-shirts advertising obscure goth or metal or psychedelic bands. It strikes me that while the auction crowd doesn't look diverse in terms of our usual markers, especially race, it is by a Martian mile the most neurologically disparate group I have ever encountered. For the first time, this strikes me as interesting. And in a curious way that I can't quite explain yet, precious.

Bidding starts with a much-coveted *Black Panther* goodie bag. Our host group of PyLadies approaches auctioneering as standup, lending a knockabout fizz to proceedings as they shoo prices through $16, $32, $64—numbers corresponding to the standard base-16 increments of computer memory. T-shirts reading "I'm not a wizard," "Hooli," "Electronic Frontier Foundation" and "Just shut up about Bitcoin" follow. Usefully, coders are well remunerated, with the US Bureau of Labor Statistics pegging domestic median pay at $84,000, so items fetch far higher prices than they would at a convention of, say, to take a random example, *writers*, although this figure masks large disparities between mobile or web developers at the low end and data scientists or site reliability engineers at the high. Bidders from the Big Tech cos, who've come with cash to spend in support of the cause, lift values further. A signed sketch of Guido goes for six hundred bucks, with a custom painted Python

Fender guitar and preassembled Raspberry Pi theremin kit still to come.

The highlight for me comes from far out of the blue, when toward the end of proceedings one of the PyLadies team presents an offering of her own. As I watch, I don't know much about Lynn Root beyond the facts that she's young, a programmer and active within PyLadies; that she retreats to the open backstage area in evident anxiety as a colleague explains the next lot's provenance. The story we hear runs as follows:

Not long ago, Lynn experienced a code burnout so severe she was forced to withdraw from the fray to recover. Sympathetic nods and grimaces of recognition at this announcement make the room appear to sway like a field of poppies. Needing something to keep her from the screen, the young Pythonista took up painting and soon felt the need to do it every day, appreciating the stillness and singularity she found at an easel. When the other women begged her to paint something for the auction, she was aghast at the thought. Her painting had meant so very much to her personally, but what could it mean to anyone else? And as the story unfolds and the room goes quiet, Lynn recoils at the back and is comforted by a friend: I can see she is shaking. Relentless encouragement persuaded her to proffer a watercolor of her cats, which is now brought to the front of the stage.

The effervescent auctioneer, a friend and fellow PyLady, knows this could backfire. She solicits a bid of sixteen dollars, to a silence you could spread on a slider. There is a further pause as people look around, Lynn looks down. And a dozen hands fly up. I breathe again. The painting won't go unsold. So the auctioneer suggests $32. Now *two* dozen hands. 64 bucks. More hands. *$128!* Leaping arms everywhere. $256. *$512!* And as the bids continue to escalate in a flurry of call and response, my own attention is drawn away from the happy auctioneer and beaming audience and

I become transfixed by a sight it would be easy to miss . . . of Lynn at the back of the stage, dissolving into tears and then floods of tears with the visceral release of a cliff crumbling into the sea—hanging on to a friend for support as the hammer falls simultaneously on *one thousand, four hundred and ten dollars* and the pain that brought her to this moment.

As a raucous cheer goes up, rising and sustaining over the crowd, I think this is one of the most beautiful things I have ever seen, not just because of Lynn's unfiltered response to such a show of affection and support, but because a room full of people she knows and doesn't know conspired to bestow this gift of empathy and catharsis for no better reason than because they could. In a curious way I feel they've given me something too, because I'll come away knowing this is a community I want to enter and understand. Not with all the computing power in the world could I have foreseen where this will ultimately take me.

Minutely Organized Particulars

To see a World in a Grain of Sand
And a Heaven in a Wild Flower,
Hold Infinity in the palm of your hand
And Eternity in an hour.

—William Blake, *Auguries of Innocence*

I've been thinking about what constitutes a great programmer. Not "great" in our standard US or British English colloquial definition of *(adj.) quite good, or at least not terrible*, but great as in canonically Great; part of the story of computing.

The question arose the other night when, plagued by post-PyCon jetlag, I took advice from an algorithmically delivered *New York Times* piece on sleeplessness, left the bedroom and settled in an armchair with a book I presumed no insomnia could defy. In the event, *Selected Writings on Computing: A Personal Perspective* by Edsger Dijkstra proved disappointing in that regard. Chagrined, I stayed up reading till dawn.

Most of the essays in *Selected Writings* are from the 1970s, by which time Dijkstra was already established as a giant in the field, and they provide a colorful window not just onto changes in our notion of programming, but onto the world programmers are called upon to program.

Who was Dijkstra? A background in theoretical physics may have explained the Dutchman's apparent mistrust of pure mathematicians. His marked resemblance to the actor Bryan Cranston as Walter White in *Breaking Bad* would have unnerved those pure mathematicians had he lived a little beyond his allotted seventy-one years, from 1930 to 2002. As expected, many of the essays in *Writings* are technical timepieces that mean little to me (and yet some tempt with titles like "A New Elephant Built from Mosquitoes Humming in Harmony"). His colorful reports on conferences and events from the time, on the other hand, are priceless, jeweled with an unusual combination of playfulness and style and an unsparing honesty that prefigures the blurty hacker culture of today. Like most great performers, Dijkstra was also a fan, but if he was in your audience you needed to be on your mettle, because he didn't suffer fools, even very clever ones.

In *Writings* is a constant sense that the words were penned at a time when everything was still up for grabs—before the PC, software microverse or social media fray were clearly defined or their trajectories set, a time when it felt natural for pioneers like him to describe algorithms not as being "written," but *discovered*, as though they're out there waiting for us like a new star or particle. He submitted to these pieces being published, he writes, partly to further his fluency in written English (being Dutch, we assume he already *speaks* it better than most native Anglophones), and you can tell he loves language, not least from his joy in learning the phrase "utterly preposterous" from a German colleague, which he converts to the acronym "UP" and finds great use for it at technical plenums. Dijkstra thinks nothing of quoting the visionary English artist and poet William Blake, whom he reveres, in the company of computing peers, nor of clashing with received thinking or reputation.

The current generation of coding doyens tends to be known for significant programs, languages or tools they've written or conceived, but Edsger Dijkstra's Greatness consisted in something else entirely, which was founded on his ability to identify and dismantle many small but critical roadblocks to hardware and software creation, without which none who followed could have made their showier, more user-focused contributions. How would we persuade two or more computers to talk to each other when they are asynchronous—meaning their processor clocks are out of step with each other? How do you create a distributed system, where there is by definition no centralized control, that is also self-stabilizing (because if you can't, there's no point in even thinking about an application like the World Wide Web)? These were among Dijkstra's questions and his heroic status rests in having found artful algorithmic answers to them, elegant as haikus.

Perhaps because digital worldviews were at stake in these early days, hackles frequently rose. Most fascinating is Dijkstra's report on a transatlantic "Communication and Computers" gathering at Newcastle upon Tyne in the northeast of England in 1973. Though characteristically effusive in praise of speakers he feels have approached their subjects with wit and (most crucially) depth, the Dutch code master's patience is also tested in ways that make for an acute cultural study. Where computing in Europe took place primarily within the academy, the starry US contingent in Newcastle were part of a fast-moving industry driven by a distinctively American mix of idealism and trust in the whims of markets—both of which Dijkstra regards with suspicion.

How did this divergence occur? Historians claim Britain forfeited its (and effectively Europe's) early leadership in computing when paternalistic mandarins chose to classify the work of pioneers like Alan Turing after World War II. By contrast, the United States, big and openhearted at its best, invited the world in and thereby took the lead. But now Dijkstra sensed danger and thought he saw scientific truth being skewed to the requirements of large corporate interests. IBM, Bell Labs and ARPA had all sent large contingents to Newcastle, and the incisive mind of Dijkstra was troubled to see good scientists dancing around them, even censoring their own research to suit commercial investment priorities. This resulted in compromised systems that, once built, couldn't easily be unbuilt and would constrain potentialities into the future. The ARPANET, forerunner of the modern internet from 1969, was one such avoidably flawed project—at a time when ARPA was trying to offload it onto the private sector, where companies like AT&T saw no future in it and passed.

Sparks flew when Dijkstra attended a pair of long talks by one of the totemic figures of American computing. His disinclination to name the speaker could look like gallantry had we been left in any doubt that the man in question was none other than Douglas Engelbart, head of the Augmentation Research Center (ARC) at Stanford University, whose team of young Californians preached the utopian vision of a networked world in which knowledge—and therefore wisdom—would be distributed; in which our electronically connected species would be held in a new embrace of fellow feeling. Wouldn't empathy arise spontaneously from such connection? How could prejudice and xenophobia resist withering away once everything was shared?

What follows is a little shocking to me. Five years prior to the Newcastle summit, Engelbart had dazzled an audience of US computing peers with a vision for the digital future. His feature-length presentation was entitled "A Research Center for Augmenting Human Intellect," and demonstrated the first hypertext links, wooden "mouse" and graphical user interface (or GUI, meaning a navigation screen that was graphical and used symbols for actions rather than text). By all accounts, most who were present in San Francisco's Civic Auditorium that day went away with their view of computing and its place in the world changed. Now Engelbart repeated his ideas for a European computing audience in Newcastle.

To those of us who saw and were excited by the democratizing, decentralizing potential of the web in the 1990s, Engelbart has always been a hero. His generous view of humanity and the "global wisdom society" was tied to a countercultural faith that people were innately collaborative and decent; that on occasion society bent this inheritance toward narcissism as wind bends a Monterey cypress—but that knowledge and connection could remedy such distortions. How much harder it should be to drop bombs on people you can see and communicate with. Except that, because the revolution faded into the recession of the 1970s, society never came to a clear verdict on whether people enter the world good and requiring only encouragement, or with chaotic hearts in need of taming. Most of our big policy debates on issues like education, the penal system and welfare continue to revolve around this irresolution. But in 1968 in San Francisco and 1973 in Newcastle upon Tyne, Engelbart's vision was radical and optimistic and most were willing to give it a try. Why wouldn't they?

Not Dijkstra. In three separate addresses to the conference by Engelbart, the Dutch master saw something not just "terribly bad," but also "dangerous." The first of these verdicts is the least shocking, because no one ever accused Douglas Engelbart of being a magnetic speaker. Neither was the second to do with substance, because "the product he [Engelbart] was selling" was "a sophisticated text editor that could be quite useful." Of graver concern was what Dijkstra saw as the "the way he appealed to mankind's lower instincts while selling . . . an undisguised appeal to anti-intellectualism [that] was frightening."

Now Dijkstra rehearses a series of complaints whose prophetic nature is only just becoming evident. Talk of "augmented knowledge workshops" reminds him of the US education system's extremely "knowledge-oriented" ethos, in opposition to his view that "one of the main objects of education is the insight that makes a lot of knowledge superfluous." A thought that puts me in mind of the coding nostrum that a good programmer distills a function down to the fewest lines, where a great one finds a way to make it redundant altogether—an approach from which most of us civilians could learn.

More striking yet was Dijkstra's instinctual (and at the time counterintuitive) dismay at Engelbart's overarching project, which was about connecting people on a global network, expressed in language foreshadowing Mark Zuckerberg's four decades later. In essence what we glimpse here is the clash of two visionary programmers of the same generation, reacting in divergent ways to the legacy of the Second World War. Where the Stanford professor was just old enough to have served in the Navy for two years—but for a country that had never been invaded—the Dutchman's childhood was defined by Nazi occupation and its assault on the individual. All the same, it is hard not to feel a little awed by the acuity of Dijkstra's projection into the future; by the speed and accuracy of his ability to unspool the logic of a proposition when he says of Engelbart, fairly or not: "His anti-individualism surfaced when he recommended his gadget as a tool for easing the cooperation between persons and groups possibly miles apart, more or less suggesting that only then are you really 'participating': no place for the solitary thinker."

Despite some contrastingly enjoyable talks, Dijkstra went away frustrated. Newcastle's highlight for him was a visit to

Hadrian's Wall, a barrier constructed across the north of England by the eponymous Roman emperor from AD 122, with "a true archaeologist!"

The point of all this for me, however, is to explain why the definition of programming Dijkstra provides in *Selected Writings* carries such weight and seems so interesting, not least because most of us feel we know what programming is by now, so never ask the question or imagine what an answer might be beyond "telling a computer what to do." But in Dijkstra's pomp, with first principles still subject to debate, the question still seems necessary. In accepting a 1974 award in the name of all the colleagues who had contributed to "the cause," he says:

> *The cause in case is the conviction that the potentialities of automatic computing equipment will only bear the fruits we look for, provided that we take the challenge of the programming task and provided that we realize that what we are called to design will get so sophisticated that Elegance is no longer a luxury, but a matter of life and death. It is in this light that we must appreciate the view of programming as a practical exercise in the effective exploitation of one's powers of Abstraction.*

Powers of *Abstraction*? A word I've noticed computerists using like a key to the digital door, but have not yet begun to grasp, whose pursuit will draw me far beyond my comfort zone, deep into the microcosmos. Dijkstra ends this speech with a quote from his beloved Blake.

> *He who would do good to another must do it in minute particulars*

General Good is the plea of the scoundrel, hypocrite and flatterer
For Art and Science cannot exist but in minutely organised particulars

One of Blake's most candescent prints depicts an idealized Sir Isaac Newton, father of modern science, sitting naked on a lichen-covered outcrop of rock as he leans forward in intense concentration, performing some geometric calculation with a pair of compasses. The Scottish sculptor Eduardo Paolozzi's three-dimensional, modernist rendering of Blake's image greets visitors to the British Library at the entrance to the courtyard every morning. But as Paolozzi knew, Blake did not intend to celebrate Newton with a masterpiece whose full title is *Newton: Personification of Man Limited by Reason*. Rather, Blake regarded the scientist's mechanistic reading of the world with disdain bordering alarm, and as the sun rises I spend a long time trying to unpack Dijkstra's use of Blake's quote here; then whether I agree with it or not. I get to wondering whether Blake, who famously saw "a world in a grain of sand" and toward the end of his life painted the soul of a flea as seen in a vision, would have liked the uncanniness of Einsteinian physics better . . . which leads me naturally to wondering how he would have viewed the digital realm. Perverse as it might seem and for reasons that will become clear, this matters more than you might think going forward.

CHAPTER

5

The Real Moriarty

Forget this world and all its troubles and if possible its multitudinous Charlatans—everything in short but the Enchantress of Numbers.

—Augusta Ada King, Countess of Lovelace

I remember wondering whether the Isaac Newton of domestic appliances, Sir James Dyson, had left a pause between his two statements deliberately. If so, his timing was good: he got me. I laughed. Then realized he wasn't joking.

Dyson's first pronouncement wasn't his own: he was referencing a quote commonly attributed to the one-time US Commissioner of Patents, Charles H. Duell. On the face of it, Duell's supposed declaration that "Everything that can be invented has been invented" qualified for the Dijkstrian category of "UP" with room to spare—even before you knew it was dated to *1899*. Victorian hubris didn't get any better, I thought (and laughed). Until Dyson also laughed and averred that, if Duell had indeed said this, he was probably right. The Victorians had invented just about everything. All we've done since is refine and embellish.

I dove into the patent literature and saw what the inventor meant. Cars, planes, telegraphs and telephones, all machines of the Industrial Revolution . . . we know these were Victorian. But the fax machine (1843, patent no. GB9745); the computer (1884, US395781–783); Frank Shuman's visionary plan, filed back in 1908, to replace coal with solar energy for the whole of Europe (GB190728130)? Many of the first automobile patents were for electric vehicles. The point being that this pattern of invention holds even for code and the imaginative leap upon which the digital revolution is founded—abstraction.

Two great Victorian figures made the vault to abstraction simultaneously, but in different contexts and with asymmetric effect. Both emerged from outside the scientific mainstream by reason of birth or gender, and both were extraordinary.

The first part of this story is familiar even to technophobes by now, though some of its more tantalizing details are less widely known.

On November 11, 1842, the British prime minister Sir Robert Peel made grudging preparations to receive one of Victorian England's most eminent intellectual figures, the mathematician and engineer Charles Babbage. 1842 had been a dismal year for the PM, haunted by a hungry working class and rising civil unrest against a backdrop of failed adventure abroad. Worse, the celebrity scientist he was about to meet wanted money, despite already having received half a million pounds—the price of two Royal Navy frigates—to build a Promethean mechanical calculator called the Difference Engine, which had so far failed to manifest. Yet in Babbage's fervid inventor's mind, the proposal he brought now was of incalculably greater moment: a sophisticated mathematical processor, *programmable*, with an internal storage system—*memory!*—giving it the effective ability to run algorithms. In return for the Exchequer's hoped-for investment, Babbage offered nothing less than the dawn of "a distinct new epoch." He called his invention the Analytical Engine and it was in all essential aspects what we now call a "computer."

The meeting went badly. Perhaps mindful of Peel's impatience, certainly stung by the doubts of detractors, Babbage began defensively and failed to explain the practical workings or public benefits of his machine. When Peel refused the request for funding, he created what modern computer programmers call a *fork*, a point at which latent futures diverge, leaving the trippy lost prospect of a first computer revolution in Victorian England, a full century before the one we're straining to accommodate now. One strand of a literary genre, steampunk fiction, is built on an imagined universe in which the inventor got his way.

Curiously, even Babbage hadn't understood the full potential of his invention. Rather, it was his close friend and confidante Ada Lovelace, daughter of the Romantic poet Lord Byron and a stellar mathematician in her own right, who foresaw the possibility of using numbers symbolically to represent and process any kind of data within the Engine—from the timbre, frequency or duration of a musical note to the letters of a word; sentence; book. To give a simple example, middle C on a piano might be assigned the number 60, with a duration of one second being represented by the code 01. Within this system, the number 6001 could symbolize the note middle C being played for one second. We could use addition to create chords or sequences of notes; multiplication to control parameters such as *amplitude* and *attack* to whatever degree of resolution and complexity our machine's memory could handle. This was a big, powerful idea.

In her writings on the Analytical Engine, Lovelace articulated the notion of software as we understand it now, using a special language—code—to translate between computers, which are digital and therefore recognize only numbers, and humans, whose reality is defined by an analog fog of thought, emotion and senses. No wonder she acclaimed the prospect of an entirely new science, which she called "the science of operations" but we now call computer science (CS). As events played out, the first university CS departments would take a further century to appear.

Over the past decade most of us have come to know Augusta Ada King, Countess of Lovelace, on first-name terms, but she has long been celebrated in computing: the US Department of Defense named their own programming language after her as far back as the 1970s. The only child born in wedlock to the mercurial Byron, she was effectively raised by a single mother who loved math and science and encouraged her daughter's prodigious talent for these voguish disciplines. In time Ada's gift would lead to friendship with Charles Babbage and allow her to see further into the potential of his Analytical Engine than even the inventor could. Despite being a woman and therefore excluded from most forms of official preferment, her genius *was* recognized at the time by figures such as the scientist-inventors Michael Faraday and Sir Charles Wheatstone, who commissioned her to write her now famous notes on the Analytical Engine. Had history favored Babbage's machine beyond the retrofuturist stylings of steampunk, wider fame would surely have followed.

Exceptional as Ada was, women of her social class being skilled in science and math was anything but unusual in eighteenth- and nineteenth-century England: each issue of the widely read *Ladies' Diary* included a rundown of the latest research and offered a prize for the solution of a sophisticated math problem, often created by internationally recognized academic professionals. More incredible is the fact that one such professional—a *Ladies' Diary* reader and contributor—turned out to be a man whose life and work entwines thrillingly with hers, but whose path to their conjunction in the annals of computing couldn't have been more different.

George Boole and Ada Lovelace were born within a fortnight of each other in November–December 1815, as dust settled across Europe after the clamor of the Napoleonic Wars. But any similarity between their lives ends there. In contrast to Ada's aristocratic status and relative material comfort, Boole was the son of a poor cobbler in the bustling county town of Lincoln, where he received almost no formal education, was working full-time by the age of sixteen and supported his entire family at twenty. Like his contemporary Charles Dickens (born 1812), Boole's cobbler father spent time in a debtors' prison, not least owing to a desperate attraction to science and learning and commensurate lack of interest in shoes. But the world of formal knowledge had no place for a man of John Boole's social standing at that time—a bitter truth that plagued his family. The forced auction through bankruptcy of a box of cherished books, rare possessions for a working class clan like the Booles, must have been crushing and brought shame deep enough for George and his sister Mary Ann to be taunted by peers in the playground.

And yet, from such beginnings George Boole conjured a thought revolution so profound that the entire computing industry would be built on it. When you activate a toaster or vacuum cleaner by means of a switch indicating "0" for "off" and "1" for "on," you unknowingly pay homage to Boole, whose innovations in logic underpin every event on a microprocessor. The mathematician's name is integral to most programming languages, which are configured to accept certain types of data. These *data types* include integers; lists; dictionaries; strings of characters, contained with quotation marks—plus a varying range of others. Of utmost importance is the *Boolean* type, a form of data object that has only two possible values: "true" or "false." In Python this binary Boolean logic is used behind the scenes in the most fundamental forms of algorithm: *conditional* (or *if*) statements and *while* loops. Both work by setting a condition (say, "*x* is greater than seven") then providing one instruction for what to do if a *Boolean test* against the condition evaluates to true, plus a separate instruction to be actioned if the test fails.

Most computerists never learn that Boole was also a philosopher and social reformer, as exercised as Dickens by the injustices of his age, not to mention an enthusiastic amateur poet and musician. The twentieth-century intellectual giant Bertrand Russell declared that "pure mathematics was discovered by Boole," which is quite a claim. The Irish mathematician Des McHale further posits Boole as one of Sir Arthur Conan Doyle's models for Professor Moriarty, Sherlock Holmes's math genius nemesis.

The real-life genius fathered five accomplished daughters, whose scions went on to make notable contributions to British and American intellectual life for generations to come. Unlike Ada, Boole wasn't thinking about computers—at least not until an intriguing encounter with Babbage toward the end of his foreshortened life. Rather, Boole was thinking about thought. So startling was the simple premise at the heart of his innovation that the wider world took the better part of a hundred years to grasp and apply it. When that moment came, the result was computing.

The cobbler John Boole, George's father, would die wreathed in the grief of thwarted ambition, neither the first nor last such casualty of the English class system. Yet père Boole's angst couldn't stop curiosity and thirst for knowledge transmitting to his children, and his frustrated dream of teaching would at least be realized by his son. Just as importantly, his wife Mary Ann Boole, herself carrying the shame of birth as a clergyman's unacknowledged daughter, contributed a love of music, literature and poetry that informed everything her children did, with the Liverpudlian poet Felicia Hemans and Byron's epochal postwar generational lament, *Childe Harold's Pilgrimage*, being favorites of her son's. By the end of his teens George had acquired fluency in ancient Greek, French, German and Italian, even choosing to correspond with friends and later write academic papers in these adopted tongues. His sister, named Mary Ann like her mother, recalls being appointed mourner at the funeral of a wild rabbit her brother tamed and cared for, at which his eulogy was delivered in self-taught Latin. If coders speak of "10Xers" in relation to their field, Boole the cobbler's son was a 10Xer in *life*.

At nineteen the future logician ran his own school and was trusted to deliver a keynote speech at the unveiling of a bust of Sir Isaac Newton. This honor would have carried special significance for him—and not just because Newton was (and remains) the foremost son of Lincolnshire. As science grew ever more visibly entangled in the daily life of nineteenth-century society, so a cult had formed around Newton, the rockstar rationalist seer of the Industrial Age, whose posthumous celebrity grew by the year. A brisk trade in icons tumbled into absurdity when a gold ring containing a supposed Newtonian tooth sold at auction for £730, equivalent to the price of a high-spec Model S Tesla today.

But if Newton and his antagonist William Blake represent two sides of a Victorian battle between analytic reason and magic, both lived simultaneously in Boole, who worshipped nature and was absorbed by faith, and who seeded logic with an abecedarian wonder that survives in our computers to this day—so I will come to believe. By all accounts a sensitive and inspiring teacher, the only recorded instance of him losing his temper was upon catching one of his pupils "bird nesting," stealing birds' nests from trees, a practice he considered unconscionably cruel.

Boole left no surviving account of how he came to the revolutionary idea upon which his principal works rest, but his wife Mary claimed it appeared in an ecstatic flash of understanding at the age of seventeen. Presuming his vision divine, he considered entering the priesthood, until a local Hebrew scholar helped him reframe the epiphany as concerning the relationship between mind and corporeal reality rather than between God and humankind. His idea was that thought might have universal underpinnings as fixed and immutable as algebra, calculus or elements of the periodic table—and might therefore be reckoned with in the same way.

Toward the end of his life, perhaps influenced by his younger, religiosceptic wife, Boole would claim to be faithless. He was certainly churchless. But faith and belief had been abiding interests up to then, and it should be no surprise that he tended against monotheistic hierarchy toward monism and a more holistic view of creation. Mary also claimed that, like her mathematician father's friend Charles Babbage, her husband was highly influenced by Hindu thought.

Through his twenties Boole built a local reputation as teacher and thinker, but wider recognition was elusive without the kind of institutional attachment a man of his status rarely saw in England. In 1847 he applied for a professorship at the new Queen's College, Cork, in Ireland, bolstering his case with an astonishing first published work, which his sister Mary Ann described as having been written in a kind of trance. Boole's book was called *The Mathematical Analysis of Logic.*

Where John Boole had been a political radical, George tended in his youth to a practical humanism, lamenting "the age of fanaticism" and regarding organized movements and institutional religion with similar disappointment. But inhumanity was everywhere in that combustible moment. At particular issue were the Corn Laws, which banned grain imports and kept bread beyond the means of workers toiling long, unrestricted hours in often dangerous conditions, with children as young as six doing likewise—while factory and farm owners fought regulation on grounds that it would undermine competitiveness and inhibit workers' freedom (in case anyone wondered where Uber et al. got that argument). With hunger and disease rife, significant movements grew around repealing the Corn Laws, restricting child labor and extending the franchise. Neither were such campaigns restricted to those directly affected: Lord Byron's first speech on ascending to the House of Lords at the age of twenty-two was in defense of Luddite stocking weavers condemned to hang for smashing machines brought in to replace them in Nottingham.

"I have been in some of the most oppressed provinces in Turkey, but never, under the most despotic of infidel governments, did I behold such squalid wretchedness as I have seen since my return, in the very heart of a Christian country," he railed.

Of dire concern to those craving change was a flood of disinformation and unreliable commentary spurred by the new steam-powered printing presses, often used to serve the status quo and discredit advocates of reform. At the heart of establishment resistance was a rigid "laissez-faire" doctrine insisting that the "invisible hand" of the market knows best and should not be fettered. As a means of justifying government inaction, this laissez-faire economics reached a peak of infamy in 1848, when Sir Charles Trevelyan, the Whig government minister in charge of relief programs in Ireland during the Great Famine of 1845–1849, allowed Irish grain to be exported rather than used to feed the starving. By famine's end the Irish population had plunged *20–25 percent* through starvation, emigration and homelessness—a calamity Trevelyan ascribed to God's righteous, Darwinian judgment, declaring the "great evil" to be not famine but "the selfish, perverse and turbulent character of the people."

Boole explicitly saw himself as following in the footsteps of Aristotle, the father of logic, whose work stood as the last word on the subject for over 2,000 years. He wasn't aware the German Enlightenment philosopher Gottfried Wilhelm Leibniz had also toyed with the notion of an "algebra of thought" (although he was later delighted to learn of it), but for an Englishman living in such volatile surroundings, the attraction of a mechanism for teasing truth or falsity out of sequences of propositions is easy to grasp. His startling insight was that the universe and everything in it could be represented using combinations of just three categories of symbol. These were:

(a) ones and zeros, to indicate *true* and *false* or *everything* and *nothing*;
(b) the basic operators *and*, *or*, *not*;
(c) "variables" to which "values" could be assigned (e.g., x = country singers named "Hank")

Boole further saw that once a premise or set of premises had been represented symbolically in this way, any conclusion contained therein could be drawn and tested for truth and consistency with algebraic precision. More powerful yet, spurious conclusions reached through laziness or cynicism could be exposed. Equations evaluating to "one" would be true, to "zero" false. End of argument.

Eagle-eyed readers may note that misinformation didn't vanish with Boole, but Boolean logic is exquisite in being at once deceptively simple and capable of infinite complexity. That his lexicon of *variables*, *values*, *operators* and *functions* is common to computer programming contains not a hint of coincidence. Variables are so called because they can be assigned any value we care to give them: *x* might be used to store the integer 186,282, or the string "186,282 represents the speed of light in miles per second," or a short list containing the wit and wisdom of Elon Musk. Here we see why programming languages specify discrete *data types* to avoid situations such as a machine being asked to divide 186,282 into the wit of Elon Musk. Most languages require programmers to manually *declare* data types, specifying *str* for string, *int* for integer and so on, letting the interpreter or compiler know what type of data it is dealing with—and helping to avoid mismatches. One of the things beginners like me appreciate about Python is that it belongs to a class of languages categorized as *dynamically typed*, where the interpreter identifies data types automatically, sparing us the trouble of declaring them in our code.

Like most symbolic languages, Boolean algebra looks (and pretty much is) intimidating at first, but the specifics of the symbols are not important here: what matters is the logic of how the world gets represented and how these representations can be operated on. Say a trio of witches hope to tempt a future king of Scotland to his doom. No problem. Witches have a potion for that, and a recipe for the potion. But what if even Aldi doesn't have all the ingredients? We might represent the resulting compromise thus:

if:

x = standard ingredients for the Weird Sisters' potion;

y = blind worm's sting;

z = owlet's wing;

then

x—(y + z) = the potion, minus blind worm's sting and owlet's wing

The Sisters know this compromised potion won't work, so they try Whole Foods, where they thrill to see blind worm's sting and owlet's wing on sale—it's Black Friday—but find eye of newt and toe of frog too overpriced to stomach. As such, we might *reassign* the variables like so:

y = eye of newt

z = toe of frog

In which case:

x—(y + z) = the potion, minus eye of newt and toe of frog

Happily, this will work. Macbeth is doomed by his own ambition, which the civic-minded witches have merely tweaked, perhaps sparing Scotland a long and brutal Macbeth dynasty. The point is that we were able to reassign the variables without changing the validity of the output: this formula (or *function*) is universal. Note again that we are interested only in logical truth; whether a statement is empirically true in the external world is of no concern to a logician or a computer.

Boole's logical vision is interesting in a multiplicity of ways. The concept of "zero" is so familiar to us now that we take it for granted, but until recently it was thought to date only from eighth- or ninth-century AD India. New carbon dating of birch bark containing the earliest known use of a symbol for zero recently floored historians and mathematicians at the Bodleian Libraries in Oxford by dating it five hundred years earlier than previously supposed. An implication of this discovery is that zero took almost a millennium to reach and be understood in Europe. The name *zero* derives from the Hindu word *sunyata*, whose closest translation is usually given as "nothingness," and scholars suggest the concept could only have arisen in a culture where nothingness—void—was embraced in a way generally alien to European thought. While Eastern mathematicians used zero (and its attendant presumption that nothing could be *something*) to extend and deepen their practice, Greek philosophers debated whether nothing could even be discussed and certainly saw no need for a corresponding symbol.

The Catholic Church later banned zero as satanic, a place devoid of God's presence, so there is a reason our word "algorithm" derives from the ninth-century Persian mathematician Al-Khwarizmi. The statement "there are no turnips in that field" is not the same as "the number of turnips in that field is zero," and appreciating this distinction made algebra, algorithms, calculus and Boolean logic possible. No wonder Fibonacci, writing at the start of the thirteenth century, gushed that "The method of the Indians surpasses all known methods to compute . . . they do their computations using nine figures and the symbol zero." One of the claims made for Boole is that he liberated zero as a bird from a cage, allowing it to fly in the Western imagination as it long had in the East.

The initial response to *The Mathematical Analysis of Logic* disappointed Boole. A few established professionals like the polymath Sir John Herschel grasped its originality and some of its implications, but most reviewers had no framework within which to fit the work, nor prism through which to see it. Resigning himself to the truth that "It will be some time before its real nature is understood," Boole decided publication had been precipitate, a pathfinder for the masterwork he would start that very same year. The full title of this work, *An Investigation of the Laws of Thought on Which are Founded the Mathematical Theories of Logic and Probabilities*, won't win any awards for pithiness, which is why mathematicians and philosophers contract it to *The Laws of Thought*. If the writing of Boole's first book was intense, its successor manifested in a kind of ecstasy. Sister Mary Ann provides a vivid description of her brother's mental state as she saw it.

> *When in 1847 the true light flashed upon him and he entered upon the investigations that resulted in* The Laws of Thought, *he was literally like a man dazzled with excess of light, his restlessness and excitement were extreme, he seemed to be dwelling in a world apart and unconscious of all that was going on around him; his countenance was lighted up with a mingled expression of delight and something like awe, and sleep and appetite deserted him. If he could have communicated his thoughts and feelings to some sympathetic mind it would have been a relief to him, but his father was gone, and there was no one near to whom he could have made himself intelligible.*

Few of us will ever know what such a state feels like, but when *The Laws of Thought* finally appeared in 1854, recognition of its importance was instant. Fortunately, *The Mathematical Analysis of Logic* had already smoothed his way to a professorship at Cork in 1849, at the tail end of the five-year Great Hunger, whence he found the streets full of homeless people while academics and church elders around him dined on turtle soup and champagne ("a more homely style of entertainment would be more suitable and in better taste," he wrote sister Mary Ann). The previous year, 1848, is known to historians as "The Year of Revolution" in Europe, and radical ideas were in the air, met with near equal force in the opposite direction: constant attack by the Catholic Church would be a feature of academic life at Cork, one of many frustrations Boole found there. Nonetheless, as the famine receded, the newly minted professor threw himself into the life of Cork, adding two groundbreaking works of pure mathematics to his contributions in logic, even if his ambition to branch into humanities research went unrealized. A happy late marriage to Mary Everest Boole at the age of forty was further sweetened by steadily growing fame and involvement in a range of educational, social and political reform movements.

For all the admiration accorded Boole in the last decade of his life, he was right to envision the rich import of his ideas taking time for the world to absorb. Only around the turn of the century did the Harvard philosopher Josiah Royce and his students give the name "Boolean algebra" to the Englishman's system of logic. Four decades later, Claude Shannon, an American master's student in electrical engineering who encountered Boolean algebra in an undergraduate philosophy course, realized that if it were possible to represent the world symbolically using binary logic on a page, then the same would be true with switches in a circuit. Drawing on the *switching circuit theory* of Japanese engineer Akira Nakashima, Shannon's schema made use of the fact that a circuit could be in one of two states at any given moment, either allowing current to pass or not. If the first state equated to one (or true) and the second to zero (or false), then electrical switches could be combined into vast arrays, generating numbers at the near-light speed of electricity—numbers that could be used to represent things in the world. Remarkably, from 1932 onward the German computer pioneer Konrad Zuse was working from the same principles. As the new techniques took off in the 1950s, Boolean logic ascended to the philosophical and mathematical mainstream, a station it retains to this day.

An ethereal innovation that illuminated the world but had no clear practical use wound up changing everything a hundred years later. When politicians discuss science and thought in terms of short-term economic gain alone, someone should force a copy of *The Mathematical Analysis of Logic* into their hands.

Perhaps the most titillating feature of this story is that it came close to being *very* different. After George's death, his wife Mary solicited recollections from his many friends and colleagues, with the aim of producing a biography. The reading of these missives and multitudes Boole wrote while alive only deepens a sense of sorrow that such a generous spirit should have been taken early (while the genocide-monger Sir Charles Trevelyan stomped on to the age of seventy-nine.) Only upon reaching a trio of long letters supplied by Boole's mathematician confidant Joseph Hill, however, does the full cost of his untimely loss become apparent.

Among Hill's many fondly told tales is a firsthand account of going with Boole to the International Exhibition, a giant world's fair reported to have drawn six million visitors to London in 1862. On perusing the tens of thousands of exhibits, Hill recalls drawing his companion's attention to a working section of Babbage's Analytical Engine, whose mechanical operation was obscure to them both. When Hill found a man discussing the Engine knowledgably with a woman friend not long afterward, he arranged to meet them again with Boole, who was at length found inspecting the Picture Gallery with Hill's sister. Recalled to the main exhibition area, the logician grew impatient over a delay in his promised instructors' arrival and had to be dissuaded by the Hills from resuming his tour of the art, until the man with whom Hill had arranged the appointment returned, followed by none other than Babbage himself, doubtless excited to meet the now renowned Boole. For some considerable time the two great mathematicians discussed the machine excitedly as a crowd formed and was even treated to a demonstration of the stunning wave-like motion of the Engine in action. Hill writes,

> *It was very interesting to see a gentleman who had devoted his whole life to the construction of a machine which should do some of the work of the human mind, explaining his admirable invention to a gentleman who had gained great celebrity by his mathematical discoveries. . . . As Boole had discovered that the brains of reasoning might be conducted by a mathematical process, and Babbage had invented a machine for the performance of mathematical work, the two great men together seemed to have taken steps towards the construction of that grand prodigy—a calculating machine.*

He concludes that, "The occasion and the scene would have formed a good subject for a painter." If only.

Much as we might wish the first, unidentified woman in Hill's account to have been Babbage's dear friend and champion Ada Lovelace, she was not: always frail of health, Ada had died ten years earlier at the age of thirty-six, same as the poet-father next to whom she is buried in the Nottinghamshire village of

Hucknall. Byron exited the world a popular hero in England and Greece, which he had been fighting to liberate from Turkey, while his daughter became the patron saint of posthumously appreciated brilliance, whose expansive "Notes on the Analytical Engine" go far beyond anything Babbage is known to have conceived—and contain a sketch for what is argued to be the first published computer program. She writes with all the luminescence of her father, too.

So, Boole and Babbage did meet, despite the latter being based at Cambridge and the former, one of the most germinal philosophers and mathematicians of his or any age, being studiously ignored by that institution (a snub recorded to have wounded him). Imagine if it had been otherwise: if Cambridge had managed to see past the class- and gender-bound mores of its time—continued in large part up to the present day—and placed Babbage, Boole and Ada Lovelace in proximity. Or would institutionalization have dimmed the outsider creative light of Lovelace and Boole, as Boole is known to have feared? We can never know, because two years after meeting Babbage, the groundbreaking logician was caught in a storm en route to a lecture and, with his health already teetering, succumbed to complications from pneumonia at the age of just forty-nine.

So: abstraction. I'm not the only American to have struggled with the idea, it seems. My favorite moment from the entertaining Netflix series *American Factory*, which chronicles the culturequake when a Chinese tycoon buys a former General Motors plant in Ohio, comes during an orientation session for Chinese workers about to meet their local counterparts.

"You will notice some differences from China," the instructor offers with magisterial understatement. "As long as you are not doing anything illegal you're free to follow your heart. You can even joke about the president: nobody will do anything to you. They don't place a heavy importance on clothes and attire. If you travel in Europe in the summertime and you see someone walking in front of you, if he's wearing shorts, vests and sports shoes, they must be an American."

The poor students look stupefied as the instructor adds, "They dislike abstraction and theory in their daily lives."

Is this true? It might be. But as Boole teaches us, abstraction in the computing sense is not so mysterious after all—we use it all the time even if we think we don't, and might frame it as a process of hiding the intermediate steps in a chain of events (the very essence of computing). If I want to keep a squirrel off my birdfeeder without harming more than its pride, I can stand vigil with a water rifle and soak the varmint as it sidles up to the seeds. Effective but tiring. Or I can train the rifle on the bird feeder with an automatic trigger wired to a small red button next to the armchair in which I sit stroking a white Persian cat and chuckling softly on behalf of bushtits everywhere . . . and if someone asked me "How did you soak the squirrel?" I would say "I pressed the red button," which would be true, but only by a process of functional abstraction. Moreover, if I introduce a form of rudimentary neural network to track the squirrel, thereby automating the process completely—at PyCon someone actually demonstrated how to do this—then the layers of abstraction start to look something like a stack. Each new layer we add to this metaphorical stack of abstraction allows us to stop thinking about the one below it and simply take its function for granted. Computerists speak of every such layer as being inside a *black box*—working but invisible and no longer requiring our attention.

Alternatively, I can learn from the very best programmers and rethink the problem, hang my feeders on unclimbable wire and scatter seeds across the forest floor to keep the fluffy rodents happy down there, allowing us to coexist and even learn to be friends in a way Boole might have appreciated. No abstraction necessary, and a pointer to the kind of coder I increasingly feel myself becoming: one who doesn't write much code.

CHAPTER

6

The New Mind Readers

I never guess. It is a capital mistake to theorize before one has data. Insensibly one begins to twist facts to suit theories, instead of theories to suit facts.

—Sherlock Holmes, *The Sign of Four*

The hotel room is small but serviceable. Sun streams through an open window.

The TV is on but I'm not paying attention because my head is full of the Python I was cramming on the ninety-minute train ride from Berlin to the German city of Magdeburg. Tomorrow morning my brain will be scanned as part of a study into how humans process computer code, which turns out to be more disquieting than I would have guessed. All week I've caught myself slipping in and out of daymares in which a neuroscientist leading her international team of researchers squints into a battery of monitors and breathes, "So *that's* where the lost socks go." Or worse, "But it appears to be made of *Play-Doh*."

A kind of inchoate unease has formed the backdrop to this trip already, the sense of a world straining to render. I'd flown to Berlin from Athens, tousled heart of a country I love but where I found the population wearied in a way I never imagined seeing, spirit worn after years of austerity brought by a financial crisis few professed to understand. If 2008 made one truth clear, it was that money—even in ostensibly physical form—is now fully virtual in essence, hard to track or weigh against conventional metrics of value like productivity or labor or social worth. Economists can no longer agree what the medium of cash *is*, much less what it does or how it might behave down the line. The only thing we can be sure of is that at the most fundamental, lowest level, it's computer code. "Just" code.

After dusty Athens, Berlin shimmers on the bright fall morning of my arrival, radiating the confidence of a city assured its place in the world. A pair of erstwhile coding neighbors had just relocated to the German capital with their previously London-based Euro division of a US tech firm, advance trickle of what many assumed would be a post-Brexit *Überschwemmung*. By stealth Berlin had become the European startup hub and a certain swagger seemed to go with that territory. Yet, as I crossed the teeming piazza outside Charlottenburg rail station, I realized I couldn't have named you a single global German or European software-based tech firm—and the question of why was starting to look interesting to me.

Of more immediate personal concern were the code "snippets" I had received the previous week, short functions of six to eight lines I would be required to interpret in the scanner, but whose surprise mathematical bent drew a nostalgic pang for my days of HTML fluffy kitten apps. I was going to be the first beginner—in fact the first non–computer scientist—to take part in the study and feared wasting my researcher hosts' time by simply going blank or not knowing enough. Would such a no-show reflect in the scans? I don't know, which is why I'm lying on the hotel bed poring over the ever more battered copy of *Python Crash Course* I've been lugging around Europe for the past month; why I am slow to register an ambient disturbance seeping through my window as if smuggled by the sun . . . a rumble of voices resolving into a chant, distant at first but drawing closer.

I rise from the bed and follow the sound to find a march of sixty to eighty mostly young people, mostly wearing black and carrying spray-painted banners I can't read because they are in German. But the banner nearest me shrieks a swastika. *Of course.* A few weeks ago, anti-immigration riots roiled the city of Chemnitz not far from here. For a moment I wonder whether these marchers are pro or antifascist, before deciding it's beside the point. I turn back to the room and reach for the comparative comfort of *Python Crash Course*, but not before catching sight of a CNN report on a noxious and dispiriting Senate hearing into the sexual conduct of a divisive potential Supreme Court judge, followed by fresh revelations about Facebook's nefarious role in the election that begat this spectacle.

And if the past few months have been restive for America as a whole, they've been yet more traumatic for tech, a congruity by no means coincidental. Where the early part of the year was all about Facebook, Google has now snatched the baton of corporate irresponsibility. Just before I left California the company came under fire for having hidden a data breach, even if data breaches were starting to seem routine. Less easily dismissed were revelations of a no-longer-secret Department of Defense contract to supply "AI" for a new category of self-guiding missile drones, which, once launched, would be beyond human control. News of a second secret project, this one to develop a search engine compliant with the surveillance and censorship aspirations of the Chinese state, prompted a tech industry first, as small but significant numbers of staff raised objections to working on such software—publicly—and a few, including a senior research scientist named Jack Poulson, felt compelled to resign. With haunting imperviousness, other company engineers followed these shocks by demoing a code robot capable of tricking a restaurant booker into thinking they were dealing with a fellow human—and of recording the conversation without disclosure. Both of which it did in the demo.

* * *

Somewhere in a distant galaxy is a spacefaring race of superintelligent beings who have worked out how to lie motionless and silent in what looks like a commercial jet engine for forty-five minutes without their noses starting to itch. To be any use to me they will need to arrive in the next ten minutes. I'm wearing a mask and my head is in a brace, earplugs inserted against the fMRI machine's unearthly clangs and groans. In my right hand is a small pad containing a single red button and in my left a ball to squeeze in case of panic. It occurs to me that this is probably the closest science gets to flying business on Aeroflot.

Not that there aren't more pressing things to worry about. Before entering the scanner I had to confirm I don't suffer from claustrophobia, despite a sneaking suspicion that I do; to empty my pockets and sign a form promising that I don't contain metal parts with the potential to make me melt or explode, which might, y'know, compromise the data—plus another form indicating my wishes should anything more concerning than odd socks or Play-Doh be found up there ("Don't worry, we wouldn't tell you, we would pass the scans to a doctor," I was reassured not altogether reassuringly). I grit my teeth as the ten-minute calibration process begins, but not too hard, because more than two millimeters' movement will turn all this effort to junk. To anyone who thinks science is easy: it's not.

I'm here, *we* are here at the Leibniz Institute for Neurobiology, one of Europe's premier centers of fMRI scanning, thanks to the grit and tenacity of Dr. Janet Siegmund. She and I had spoken on the phone beforehand, but over a pre-scan coffee with two other members of her international team—a Polish doctoral candidate named Norman Peitek and the neuroscientist André Brechmann—she talked me through the start-stop history of scientific research into how code is represented and processed in programmers' minds, and why this matters.

Academic psychologists began to research coding's effect on the brain in the 1970s, but never reached consensus on how the brain's treatment of one of its newest inputs could be assessed. The most memorable intervention came from Dr. Philip Zimbardo, famed author of the Stanford Prison Experiment and seminal book *The Lucifer Effect: Understanding How Good People Turn Evil*, who in 1980 published a study in the journal *Psychology Today* titled "Computer Addiction: Reflections on a New Obsession." The article described a Stanford computer science cohort who "were losing touch with the human race"—and knew it. Zimbardo feared that predominant features of "hackerdom," not least the mulish pursuit of narrow goals and devaluation of people and relationships, were becoming general to American society, infecting us all. Nonetheless, finding it hard to pinpoint a mechanism by which the process of coding could be examined, most psychologists gave up and left the field until the arrival of Big Code in the mid-2000s. Now researchers could mine vast data repositories such as GitHub for statistical clues as to what coders did and how different language constructions and paradigms helped or hindered them. But such material offered only inferences based on available choices; it revealed nothing of what was happening within a coder's mind.

All of which counts for reasons any code instructor worth their salt understands. One of the big issues in programming education is the astronomic dropout rate, held to be a brake on gender, racial and ethnic, cognitive, economic and cultural diversity, beyond the already substantial challenge of access to resources. A Python in Education Summit prior to PyCon each year finds teachers, podcasters, writers of instructional books and public servants agonizing over this phenomenon and pooling ideas for how to tackle it. Working on trial and error or hunch, theirs is an uphill slog.

Programmers have long dreamt of a language that could be natural and intuitive to humans while still addressing a microprocessor efficiently. But the way classical computers and minds process information is so radically different that a "semantic gap" emerges. The problem is that this semantic gap moves reciprocally to the stack of abstraction: narrowing it with a more human-accessible programming language can only be achieved by adding layers of abstraction, thereby increasing the stack, which—as we know—eats up efficiency, reliability and control. For the foreseeable future, then, Dr. Janet Siegmund's semantic gap will remain, but up to now there has been almost no research-based information on how best to manage it. Language designers and teachers alike fly blind. This baffling chasm in our knowledge is what Siegmund's team has begun to address.

The one thing everyone agrees upon is that the traditional programming curriculum in schools has been ineffective to the point of counterproductivity and that the dropout rate even among motivated mature students is immense. With so little information to go on, most of us assume computer code is like the algebra it superficially resembles, which gives an advantage to the overwhelmingly white and Asian male science and math specialists who fill the university computer science departments that feed Silicon Valley. I haven't yet seen enough to know why this status quo endures, but I do know some Valley recruiters question the value not just of coding bootcamps, but of the supposed gold standard, CS degrees, which suggests a systemic problem of some kind. How could so little be known about an activity of such urgent and accelerating consequence?

As the team spoke, I learned how. By the time Dr. Siegmund had completed her degrees in psychology and computer science in the first decade of the twenty-first century, a range of new imaging technologies had revolutionized scientists' understanding of brain function, to the point where the tasks associated with different regions were well understood. The most accessible of these technologies, functional magnetic resonance imaging, or *fMRI*, works by tracking blood flow to the various neural regions, believed to be a reliable indicator of activity. As such, it is a blunt tool whose limits are carefully acknowledged by neuroscientists, but it can be very effective in the right contexts.

Siegmund saw the mysteries around programming as one such context. The trouble was, everyone who heard her idea said she was dreaming, this could never work—and as I lie in the scanner pondering the related mystery of how the hell I got here, I feel specially positioned to appreciate why. For a start, neither the keyboard nor monitor exists whose electrics won't fry in an fMRI machine. So programming *per se* was not immediately possible. Yet, like Guido van Rossum before her, Siegmund understood that code is read more than written: if a way could be found to get visual data into the scanner, a study of code *comprehension* would have great value, both in itself and as proof of concept. At a conference in 2006 Siegmund met André Brechmann, head of Special Lab Non-Invasive Brain Imaging at the Leibniz Institute, from whom I learn more about brain science in a couple of hours than from everything I've read on the subject previously. While Brechmann agreed with peers that designing such a study would be difficult and maybe impossible, he didn't question Siegmnund's sanity.

This was all she needed. A technique for relaying code snippets into the fMRI machine with a small mirror was developed. Typical snippets would be short functions for which an input value was provided, leaving the research subject to calculate a corresponding output. Sometimes there would be red herrings in the form of deliberate syntactic mistakes. A subject then had up to thirty seconds to compute the output or identify an anomaly and press a button to move on. How to make experienced programmers work hard enough on such short snippets to produce meaningful data without resorting to counterproductive esoterica, though? Easier said than done.

Developing a viable process took the better part of two years. Over that time a routine evolved through which code snippets alternated with blank "rest" and separate "d" screens containing a very simple, puzzle-like task designed to clear the subject's head. In my case there was the extra challenge of converting Java, in which prior CS undergrad participants were all fluent, into my beginner's Python. The doctoral candidate Norman Peitek inherited this task, made trickier by his having no way of knowing how long the snippets might take me to figure out—or even if I would be *able* to figure them out. Little wonder he and I have been in constant communication these past few weeks and greeted each other like old friends when he picked me up from the hotel this morning.

I feel calm and focused as a first "d" screen appears, less so with the arrival of the first snippet, which is whipped away before I've managed to calculate its output, beginning a steady cycle of work—rest—work—rest—work. Claustrophobia may not be an issue, but as the code flies past I'm struck by a curious feeling of vulnerability—almost of being *seen* for the first time. Which in a sense is true. I try not to react defensively by hitting the button when I haven't finished my calculations, and for the most part do, reminding myself this isn't a test and the most important thing is the work going on in my head. Funny that I should feel an impulse to hit the button anyway. Am I really that vain? Can the scanner see this question? Not infrequently the thirty seconds are up before I'm done with a function, and of course when red herrings appear in the form of inappropriately named methods or suspect syntax, my brain has a whole extra set of questions to ask than would a code-native counterpart, not least "Is this wrong or have I just not learned it yet?" In this way time passes in a flash and I'm hauled from the machine with almost no recall of what occurred. A chance to review the code and my response to it will come when the results are collated and analyzed—a big job, which could take months given the workload of the team. Until then neither I nor the people in the control room have any idea how my brain responded or what its response might mean. I stand and exhaustion flushes through me like a hoard of tiny bots.

The most distinctive structure in Magdeburg looms over one side of the central Breiter Weg. In keeping with the philosophy of its brilliant Austrian-born artist-designer Friedensreich Hundertwasser, the Grüne Zitadelle (Green Citadel), incorporating the ArtHOTEL Magdeburg, contains no straight lines and looks like something you would eat more likely than stay in for under fifty bucks a night. You can tell if someone is seeing it for the first time, because almost without exception their faces light up like a child's. Hundertwasser's phantasmagoric creation communicates nothing explicit yet radiates fun and joy and a vision for the future; it suggests new ways of seeing just by being what it is. For the artist, straight lines were "godless and immoral." He didn't use them.

Warmed by low sun and with mind freed after the tension of the scan, I use the stroll back to the ArtHOTEL to review what I hope to learn from the fMRI process. And I suppose that, like Hundertwasser, my concern right now is with straight lines. If coders like clearly defined boundaries between things, as everyone tells me they do, it's because the machine demands precision and strict demarcation, not because they suit an analog realm where clear distinctions are rare and tend to come at high cost ("Nature never draws a line without smudging it," Winston Churchill once noted). Is it possible to spend so much time at the human-machine interface, programming, that one day you wake up to find you're no longer a person in the fullest sense? In her classic accounting of the trade, *Close to the Machine: Technophilia and Its Discontents*, the code doyenne Ellen Ullman implies as much.

"Only a cadre was supposed to go through this," she explained to *Salon*. "The rest of humanity wasn't. Eventually, you realize, if the world is being remade by these people who've suppressed all these other parts of themselves, when they're done with all their decades and decades of struggle, will they remember how to be a complicated human being?"

And yet the most distinctive feature of the community I encountered at PyCon is radical neurologic diversity. I reflect on the "d" screens, which have a tale to tell about the relationship between brain and computer. One of the qualities that make brains hard to study is the staggering amount of background noise and activity that sustains even at rest. Norman Peitek explained that his team must filter an enormous amount out before scanning can begin. Ask why so much noise should be there in the first place and things start to get colorful.

Our digital computers are often referred to as "deterministic," meaning a particular input should always produce the same output whether passed to the machine two times or two million times. Neuroscientists are still trying to make sense of the radical extent to which this is not true of the brain. Where identical input produces the same output in a digital environment—computers would be of limited use to us otherwise—our brains *never* produce identical output in response to the same input. The question is whether this variability arises by evolutionary accident or design.

What we do know is this: the elementary unit of transmission in the brain, the neuron, has binary output, in that it fires or doesn't fire (not unlike a switch), but its input is analog; can be *graded* the way a piano string's input is graded by the force with which its corresponding key is struck. Not only that, but when a neuron fires a signal along the cable-like axon that connects to an adjoining neuron, the signal reaches that neighbor neuron at best 50 percent of the time. This seeming inefficiency is not something Apple engineers would design into a computer, and researchers can only speculate why it's there. Furthermore, at any given moment an active region of the brain is receiving information not just from an outside stimulus, via the senses, but from other regions of the brain, implying that no stimulus can ever be experienced exactly the same way twice. At the Frankfurt Institute for Advanced Studies, Prof. Dr. Matthias Kaschube and his international team are trying to understand this systemic imprecision.

"There's debate about whether the fluctuations that we see in experiments are really meaningful and contain information that we don't understand yet, or whether they're just noise arising from the stochasticity of biochemical processes and are something that the brain needs to ignore or average out," he told the science magazine *Nautilus* in a piece entitled "Why the Brain Is So Noisy."

"Stochasticity" is the property of learning by mimicry. Kaschube continues,

> *Having slightly different responses to the same stimulus can help us detect different aspects of the scene . . . it could be that the brain adds noise in order to sample different representations of what's out there. And by exploring the space of potential representations, the brain may try to find the one that is most suited given the current context.*

Wow. The rest screens in my scan are designed to mask this background noise by offering a small amount of foreground work to focus on. Asked whether the deep learning neural networks behind some forms of "AI" are wired like the brain, Dr. Kaschube explains that the connections in neural networks are typically "feed-forward," meaning input data passes through a series of processing layers before reaching an output. And crucially, "There are no loops in feed-forward networks. . . . A feed-forward network is really a crude oversimplification and very distinct from the highly interconnected networks in the brain."

Kaschube adds that,

> *A large part of neuronal activity is ongoing crosstalk among different brain areas, and sensory input sometimes appears to only play a modulatory role in this internal activity. That's a very different perspective than the one that you usually have in deep neural networks, in which neurons only get activated when they are provided with input.*

By this account the brain operates as a kind of continuum and is anything but linear. For the time being, computers and brains remain glaringly different. One of the things I hope these scans will help me assess is whether computers are on their way to matching brains, or whether through constant interaction and adaptation both are on a path to meeting in the middle, coming to resemble each other by a process of equalization.

My excitement at being part of the Siegmund team's project stems in part from surprising early results. In a post-scan debrief, I learn that in most study subjects five regions of the brain dominate the effort to comprehend computer code; that these "activated" regions are broadly associated with working memory, problem-solving and *language processing*. All are in the left hemisphere for most right-handed people, which—with a bikeshedful of caveats—nearly always takes the lead in dealing with the application of natural language (and analytical processes in general). To the team's surprise, they observed no left hemisphere activity associated with mathematical thinking, and no significant right hemisphere activation at all.

Perhaps these indications shouldn't be surprising: Siegmund quotes no less an authority than Dijkstra as saying, "An exceptionally good mastery of one's native tongue is the most vital asset of a competent programmer." She and her team appear to have found provisional evidence that the great multilingual Dutchman was right, suggesting the preponderance of people with language skills at PyCon was no coincidence. If confirmed over time and by others, such a finding would have material implications for teaching and recruitment, and might suggest ways to diversify an alarmingly homogeneous profession. As project programming lead Prof. Dr. Sven Apel, chair of software engineering at Saarland University, tells me:

"In our curriculum, we need to also educate students in English and German, because at the moment we don't do it. We should work with them more on writing text and prose, like papers and presentations. Sometimes students . . ." he pauses, searching for a delicate way to say what he's about to say. "I don't know if you know this, Andrew, right, but some computer science students are hardly able to talk, to form whole sentences when they come to us."

Apel goes on to discuss the relevance of this study to programming language design, noting that languages iterate differently, requiring data types to be declared or not and adopting multifarious approaches to syntax—all based on little concrete knowledge of what works best and worst for people.

"Often the choices will be about speed. In other words, they are optimized to the computer, with little thought about how hard or easy they might be to learn. And of course languages are written by experts. So that's why I find this research very interesting. It can really change the world. Many, many experiments are necessary, but these results have profound implications for learning and teaching."

The work so far pertains to comprehending code: as already noted, how the brain composes it is harder to assess and could be very different. But at the very least Janet Siegmund and team's research opens the scientific account for code with an encouraging dash of intrigue. Long-term goals include homing in on which languages or programming paradigms best facilitate learning and usage; on identifying the differences in brain activation between novices and experts; on helping us to understand our friend the 10xer.

"The saying is that you can't train them," Siegmund says of the latter. "All you can do is find them and let them loose. But we don't know anything about this, whether they're born or made.

Maybe we can teach anyone to be such a good programmer if we do it in the right way."

Perhaps loathe to upend even a questionable incumbent curriculum without an established body of evidence, teaching professionals have been slow to acknowledge the Siegmund team's findings. One who has taken note is a high school teacher named Scott Portnoff. After devouring a series of papers he wrote with titles like "The Introductory Computer Programming Course Is First and Foremost a LANGUAGE Course" and "Teaching HS Computer Science as if the Rest of the World Existed," I track him to Los Angeles, where he recounts decades of struggle to keep his students learning and engaged with traditional "problem-based" programming pedagogy, in which students are tasked with a problem and asked to work through the logic of how to solve it for themselves. By instinct this is how I've tried to proceed up to now, in the belief that while solutions to individual problems may not be transferrable to others, their underlying logic is. And though understanding logic is never bad, at length Portnoff grew to believe it was a poor first step. Basing a curriculum on it was "just nuts."

"The kids just weren't learning the curriculum," he tells me. "I would end up telling them, 'We don't know how to teach this.'"

As with other committed code educators I meet, he began to experiment and adapt—and implicitly to adopt methods common to natural language acquisition.

"I thought the problem was just the kids weren't interested, so I found ways to make it interesting. Now they were interested, but they still weren't learning how to program. One year, at the beginning of the second semester, there were still kids who could not write a simple program without making syntax errors. So, my job was running all over the classroom trying to put out these fires. It was so demoralizing that I just said, listen, this is what we're gonna do: here's a short program, I want everybody to memorize it. And my thought bubble says, 'I can't believe I'm asking them to do this!' I can see their thought bubbles, too, going, 'I can't believe he's asking this!' I mean, memorization—what's that got to do with programming, you know?"

Despite the incredulity of everyone involved, Portnoff stuck to his guns. Homework that weekend was to memorize the assigned program well enough to write it out perfectly. Those who mastered the task got an *A* and those who didn't got an *F* and had to spend lunchtimes practicing.

"Some kids took it seriously. And, you know, they got it perfect. And then I had some kids who were holdouts, they just dug their heels in. But I was adamant, because I saw that the kids who did it suddenly stopped making syntax errors. Once they got over this initial hump, they could write a program that would run. And I have a minor in linguistics, have always been interested in languages. So, I realized this is how we acquire language."

True. We don't teach children the etymology of a word or its full panoply of meanings before allowing them to use it. Kids listen and see how other people deploy the word, then mimic, adding conceptual depth over time. And in big-picture terms this is a beautiful thing. I don't need to know why JavaScript demands a semicolon at the end of every statement; all I need to grasp is *it does* and that computers, like bears and traffic cops, don't negotiate. As Portnoff points out, modern second language acquisition theory explicitly employs the principle of "implicit repetitive exposure to language data in meaningful contexts." Language teaching pedagogy *is* incorporated into traditional programming instruction, he adds, but modeled after an older *Prescriptive Linguistics* approach based on "explicit rule-based grammar instruction." I learned French at school by the latter method and loathed every second of it, then spent time in France and adored the process of listening and learning how to use what I heard. What Portnoff says makes sense.

The funny thing is that while I understand Portnoff's theory, it still takes a long time to trust in my own case. A voice in my head demands to know *how* and *why* something just worked, refusing to believe this won't be important down the line. But as the teacher points out, the detail of a solution to one problem in computing is often not transferrable to another. And when I make myself try Portnoff's methods, finding programs written by experts and learning to write them down on a piece of paper and then my editor until committed to memory, they help. Nuance will be essential to grasp in the end but doesn't need to be the *first* thing to know. And to my surprise deeper understanding often arises spontaneously from repetition. Better still, in aping the experts I am discovering the best way (or at least *one* good way) to do something, rather than what I'm capable of conceiving myself right now. It's hard to overstate how counterintuitive this is to me.

Of course, as Gerald Weinberg (author of *The Psychology of Computer Programming*) pointed out to me early on, an approach that works for one person won't necessarily work for another. If programmers have opened my eyes to one thing, it's the illimitable, galactic *sweep* of the spectrum on which human minds reside. I will end up adopting both problem-solving and rote approaches. Unlikely as it seems, when logic starts to overwhelm, repetition often supplies a way forward. And yet there is no panacea here: code continues to be plaited with frustration for me at this early stage. Will this change? I don't know.

The plan is to come back for a second scan if and when I'm more proficient, to see if my brain functions differently in detectable ways. Janet Siegmund has worked hard to manage my expectations, giving me the book *The New Mind Readers* by the Stanford Professor Russell A. Poldrack, which lays bare the strengths and limitations of fMRI technology. She reminds me that brains are now understood to be highly "plastic," capable of spontaneously reorganizing to meet new challenges, meaning that *any* new skill we acquire involves a degree of reconfiguration. A classic case involves London cabbies, whose traditional two-to-four-year rite of "doing The Knowledge" (learning every street and significant building or business in the sprawling city by heart) causes the hippocampus of the brain to expand until their brains are identifiably different from those of the general population. Why wouldn't some equivalent expansion, or contraction, happen in coders?

I return to the United States knowing just one thing for sure: coding will change my brain. The question of *how* turns out to be more engrossing than I could possibly have imagined.

CHAPTER

7

Theories of Memory

Surround yourself with human beings, my dear. They are easier to fight for than principles." He laughed. "But don't let me down and become human yourself. We would lose such a wonderful machine.

—Ian Fleming, *Casino Royale*

As war loomed in 1938, British planners took a pre-emptive decision to move the Government Code and Cypher School out of a London they knew would come under attack. The site they chose was Bletchley Park, a Victorian manor and estate half an hour north of the capital, equidistant the deep intellectual pools of Oxford and Cambridge. When hostilities broke out, slowly at first and then in a terrifying scrabble of invasion-retreat-retrenchment-defense, the world had yet to see a working computer. By the fall of Berlin six years later, humankind's first electronically driven conflict had forced the development of computing machines on both sides of the Atlantic, first at Bletchley and the University of Pennsylvania, then the University of Manchester and the Institute for Advanced Study near Princeton. In Germany, Konrad Zuse was on a parallel track. Through the fog of war, choices were made that would define this newest human realm for at least another century, refashioning society far beyond that.

The sky is blue. It's a cold, crisp day early in the new year and this is my first visit to Bletchley Park. Breath steams and even gloved fingers tingle, but a steady stream of visitors has come nonetheless, moving slowly around the recently restored site as if it were a cathedral; as though sudden movement or loud speech might break some lingering spell and undo the work of the brilliant minds still haunting the Gothic manor and drab huts, among whose spirits is the father of modern computing himself, Alan Turing.

Leafy and arranged around a lake Cypher School staff skated in winter, with trees planted to throw shade confusing to a German air crew, the estate is a model of green English pleasantness and I am startled by how alive it still feels—how present its history. Equally surprising is that I find myself roving the grounds feeling perpetually on the verge of tears, without quite knowing why.

Codebreakers were called together at the outbreak of war with the coded message "Auntie Flo is not so well." There were only two hundred staff then, but by 1945 there were *ten thousand*, three-quarters of them women. Nothing about the existence of this place was inevitable: the Cambridge mathematician Gordon Welchman first had the idea to reinvent codebreaking as a structured, factory-like process, industrializing flashes of intuition and genius into intelligence that could direct the entire Allied effort. The technology needed to create such a mechanism was speculative in the beginning, a phantasmal dream of Turing's that the Chiefs of Staff struggled to understand and were loath to fund. Only when Turing's ideas were smuggled to Downing Street and the acute intellect of Churchill were the necessary resources freed. "*Action this day*," he ordered. And so it was.

After Norway and Denmark were overrun in April 1940, work at Bletchley reached a pitch of intensity it would maintain for the next five years, with staff organized into eight-hour, round-the-clock shifts. At the project's peak, forty motorcycle couriers delivered up to three thousand messages a day. The writer and actor Stephen Fry once said, "Just saying the name Bletchley Park gives me goosebumps," and, standing at the cottage within which Alan Turing and the Cambridge classicist Dilly Knox made the first breaks into code generated by the wily German Enigma encryption device, or gawping at a reproduction of the beautiful "Bombe" machine Turing and Welchman designed to automate the decryption process, with its rows of revolving multicolored drums, I suddenly feel I know what he meant.

The Electromechanical Enigma machine was far from the only concern at Bletchley, but it began as the most immediate. Three rotors were used to scramble messages, generating a reported 103,325,660,891,587,134,000,000 possible combinations, with settings changed every day. Turing and Knox learned to break the code by hand, but the process took too long for most of the intelligence gained this way to be useful. Turing's "Bombe" brought the decryption process down to hours, until the German Navy added a fourth rotor to its own version of Enigma and the Bletchley codebreakers found themselves locked out for ten appalling months—with Allied shipping being picked off by "wolf packs" of German U-boats in the North Atlantic, threatening to starve Britain into submission. "[That] was a grim time," said Shaun Wylie, the mathematician and polymath who worked with Turing's Enigma team in Hut 8. "We realized that our work meant lives and it ceased to be fun."

Breaking Enigma was child's play compared to the impenetrable Lorenz teletype machine used to convey messages between Hitler, his High Command and German Army Field Marshals. The British nicknamed Lorenz "Tunny," for "Tunafish," a reference to their characterization of intercepted messages as "fish," and it was vastly more difficult to penetrate. Where Enigma resembled a large typewriter, Lorenz was the size of a modern laser printer and made use of twelve rotors. Because no Allied eyes had seen Tunny, its workings and logic were unknowable—until, in August 1941, an operator made a mistake in a message sent between Vienna and occupied Athens, equivalent to sending a small amount of metadata along with the encoded message. From this, an army cryptologist named John Tiltman, using a method developed by Turing and referred to by staff as *Turingery*, managed to break the code over ten feverish days. And from this tiny chink in Tunny's armor, the chemist-mathematician Bill Tutte was able to infer its mechanism and develop a brilliant statistical decryption logic that prefigured the modern search engine. In the course of this work, Tutte would make profound advances in the field of *graph theory*, elevating it from a mathematical backwater to a central concern of computing, with application to everything from linguistics to the natural and social sciences.

For a time, Tunny intercepts were painstakingly decrypted and translated by the skilled, largely female "computers" of a special section under the supervision of the Cambridge mathematician Max Newman. But by early 1943, German cryptographers had refined their machine to the point where decryption by hand and brain was too slow to be viable. Bletchley codebreakers realized the only way to recover the Allied edge in intelligence was to do something no one, to their knowledge, had done before. Between the wars, a series of lectures by Max Newman had inspired the twenty-four-year-old Alan Turing to conceive the possibility of a "universal computing machine." Now the desperate Bletchley team needed such a machine, or something like it, to reinfiltrate Hitler's code. They were on entirely novel ground.

🐞 🐞 🐞

While the British grappled code, US-based mathematicians addressed problems of a different kind. Munitions designers were planning and building ever more powerful weapons, whose trajectories and blast characteristics needed to be understood in order to be usable. But some of the calculations involved in modeling these operational characteristics were complex: even in large teams, human computers were coming under strain. Then came the nuclear weapons of the top-secret Manhattan Project.

The story of the first computers—and computer programmers—tends to be told like most technological narratives: as a procession of inevitable steps taken by a species of brilliant people, usually men. Draw close enough to be able to describe it, though, and you find your view progressively obscured by war and the mists of time, as even simple truths you thought you knew collapse around you. I thought the prehistory of programming was going to be straightforward. But it's not. For better and worse, two giant reputations tend to dominate the record and further befog the view.

Alan Turing and John von Neumann were child math prodigies with connections to aristocracy, the former in England, the latter as part of an affluent secular Jewish family in Budapest, Hungary. Both were also among that rarefied class of thinker given to conjuring not just theories, but whole *disciplines* with the seeming ease of potters pulling vases from a kiln. By the time von Neumann accepted a post at Princeton in 1929, then a lifetime professorship at the related but independent Institute for Advanced Study (IAS), he'd been credited with inventing quantum logic, game theory and continuous geometry, and with defining a rich mathematical framework for quantum mechanics. When Turing went to Princeton for his PhD in 1936 at the age of twenty-four, von Neumann, nine years his senior, invited him to stay on as a postdoctoral assistant at the IAS afterward. With war approaching, the Englishman instead returned to Cambridge upon receiving his doctorate in 1938, where he famously clashed with Wittgenstein during the latter's lectures on the foundation of mathematics.

Unlike the monastic Turing, a keen long-distance runner who spoke haltingly, as if speech couldn't compete with the riot of synapses firing in his brain, von Neumann is remembered as sociable, with an earthy sense of humor; a *bon vivant* who did his best work against a backdrop of noise and motion, whether at cocktail parties or rail stations or among hordes of shrieking children. For a man of reason and logic, he had what colleagues took to be a confounding superstitious streak. "A drawer could not be opened unless it was pushed in and out seven times," his wife Klára wrote in an unpublished memoir of her husband. "The same with a light switch which also had to be flipped seven times before you could let it stay." Almost a century later, these behaviors, along with his ability to perform elaborate mathematical calculations in his head, might be seen as markers for obsessive-compulsive disorder or autism. His mother reported being woken from a daydream on one occasion by her young son asking, "What are you calculating, Mommy?"

Universal among friends and colleagues was awe at the lucidity with which von Neumann could break down any idea or problem, ready to be explained in a way that left listeners wondering how the issue had ever seemed problematic in the first place.

The science historian George Dyson, in his monumental study of the creation of the first computers, *Turing's Cathedral*, quotes von Neumann's fellow Hungarian American mathematician Paul Halmos as saying, "We can all think clearly, more or less, some of the time. But von Neumann's clarity of thought was orders of magnitude greater than that of most of us, all the time." It was a distinctively logical form of intelligence, Halmos added, noting that his countryman "admired, perhaps envied people who had the complementary qualities—the flashes of irrational intuition that sometimes change the direction of scientific progress."

Such a person was Alan Turing.

When Von Neumann asked Turing to be his postdoc assistant at Princeton, he had in mind a paper the Englishman published in *Proceedings of the London Mathematical Society* at the end of 1936. *On Computable Numbers, with an Application to the Entscheidungsproblem* drew on the work of sixteenth- and seventeenth-century thinkers including Liebniz, Francis Bacon and Thomas Hobbes to address problems posed by twentieth-century German mathematician-philosophers David Hilbert and Kurt Gödel. In the course of addressing these arcane theoretical issues, Turing postulated a machine upon which any mathematical operation that was calculable could be carried out algorithmically. He called his invention a "universal computing machine," though it has since become known as the "Universal Turing Machine."

The thirty-five pages of *On Computable Numbers* are certainly among the most momentous studies of computing ever published, describing as they do how a Turing Machine could be structured and run. Up to that point, the functioning of most machines depended on the mechanics of hardware: a device could do only the things its mechanism or circuitry was engineered or hardwired to do, meaning that once its operational parameters had been set they could not easily be changed. Now Turing showed how data and instructions about what to do with the data—*software*, in other words—could share the same space in a computing machine's memory. The word "universal" indicated that, by the method Turing described, one piece of hardware could contain any number or type of differently configured machines. In other words, once the instructions were encoded in electronic memory rather than hardwired into circuitry or mechanics, they could be changed at will, by means of what we now call a *program*. The architecture Turing posited is called "stored program" architecture and describes almost all computers we use today.

Programmability was not an entirely new idea: Ada Lovelace had envisaged it. The automated Jacquard Loom of the early nineteenth century employed *punch cards*—cards into which patterns of holes had been punched—to weave complex patterns in silk. Babbage borrowed Jacquard's idea for his Analytical Engine, as did some early computers, and Ada was beguiled by the metaphoric connection she saw between these machines, writing, "We may say that the Analytical Engine weaves algebraic patterns just as the Jacquard Loom weaves flowers and leaves." But in showing that "It is possible to invent a single machine which can be used to compute any conceivable sequence," Turing's rarefied version of programmability was a giant theoretical leap. Between the lines was a genesis of sorts, a rift in the nature of time as experienced by humans and by Turing Machines. Where time to humans is a continuum and for all practical purposes *constant*, our notion of temporality means nothing at all to a Turing Machine, for which time is not constant, but stepped, defined by a linear sequence of changes of state. One significant consequence of this fork in the conception of time is that as computers get faster, microcosmic time *actually speeds up* in relation to human time. And from the perspective of the machines, we slow down.

None of which robs the Turing Machine of physical presence, because with each state change, data and instructions are stored as numbers in physical locations, most obviously numeric *addresses* within the computer's memory (Max Newman later referred to these as "houses"). Indeed, this is what the machine's changes of state *are*. Do something, store the result: do something, store the result: do something, store the result. The computer's action appears fluid to us because each change of state happens very quickly, like the blurring of frames in a film. But to a machine these are discrete, irreducible steps and a computer is no more than this sequence of steps, each involving a tiny change, in the form of a single calculation that cascades through the system.

Summing up the imaginative leap Turing made at the age of twenty-four, George Dyson states:

> *The title* On Computable Numbers *(rather than "On Computable Functions") signaled a fundamental shift. Before Turing, things were done to numbers. After Turing, numbers began doing things. By showing that a machine could be encoded as a number, and a number decoded as a machine,* On Computable Numbers *led to numbers (now called "software") that were "computable" in a way that was entirely new.*

If we think back to Professor Kaschube's postulations about the mysteries of background noise in the brain, it becomes clear that mind and computer have almost nothing in common as phenomena. Which may explain the feeling I've had from the start, not of adapting my brain to the microcosmos, but of building a new one to run parallel with the original, a kind of biological Turing machine in my head.

Contrary to common assumption—encouraged by the dissembling movie *The Imitation Game*—the electronic marvel built to decipher messages sent between Tunnies was neither designed by Alan Turing nor built at Bletchley Park. A first attempt to automate the codebreaking was of Max Newman's design and nicknamed "Heath Robinson" (after the British equivalent of Rube Goldberg) for its thrown-together look. Heath Robinson's mechanism was clever, involving two looped paper tapes, the first containing a punched, coded version of the message to be decrypted and the second a possible configuration of the tunny keys. The machine used optical readers to analyze each tape and look for patterns of correspondence, with 150 electronic thermionic valves or *vacuum tubes* operating as switches equivalent to the transistors on a modern microprocessor, organized via Boolean logic. Innovative though the Robinson system was, keeping two tapes in perfect synchronization was arduous and the tubes were perpetually failing: Newman's machine was unreliable and still too slow.

Turing persuaded Newman to seek help from a virtuoso post office electrical engineer named Tommy Flowers, with whom he had worked on the Bombe. But instead of offering refinements to a machine he thought fatally flawed, Flowers, the London-born son of a bricklayer, proposed a far more advanced machine built around 1,500 rather than 150 vacuum tube switches. Major Bletchley figures like the academic mathematician Gordon Welchman, an alumnus of the upper-crust Marlborough College school, dismissed Flowers's agglomeration of tubes as expensive and inherently erratic, favoring arrays of slower, cheaper electromechanical switches. Newman and Turing remained open to Flowers's concept, but the weight of opinion was against them. As events played out, Turing was away in the United States from November 1942 through March 1943 and appears to have had little direct involvement in the first computer's development.

What Flowers knew from his innovative research into the first tube-based digital telephone exchanges in the 1930s was that this technology could be lightning fast and highly dependable if the units were left to run constantly rather than being switched on and off. Fortunately for humankind, Flowers's boss at the Post Office Research Station in Dollis Hill, North London, had faith in his engineer. The inventor was given a dedicated fifty-person team and first call on whatever supplies he needed. None of which spared him devoting a considerable sum of his own money to the project. Flowers would tell an interviewer three decades later that, "I don't think they [the Bletchley Park academics] really understood what I was saying in detail. I am sure they didn't, because when the first machine was constructed and working, they obviously were taken aback—they just couldn't believe it."

Newman and Turing appear to have steered Flowers toward making his machine programmable and animating the circuits with Boolean logic. With no internal memory, input was via a single loop of paper tape. While the tape ran at thirty miles an hour, an optical reader took in and analyzed the scrambled text at a rate of five thousand characters per second, over and over, searching for faint patterns or eddies of design, ripples of sense in an otherwise flat numeric ocean. Five parallel processors were programmed from among a range of specialized Boolean functions—algorithms—selected using a switchboard to the machine's front, with customized functions programmable via a plugboard to the rear. With each pass of the tape, the Boolean functions would evaluate to True or False (one or zero) and send results to counters, allowing barely perceptible gradients to be detected over time until the position of all twelve daily changed rotors was known. In this way, Tunny intercepts could be decrypted in hours rather than days or weeks.

Flowers's team took eleven months to build what is agreed to be the first operational programmable digital computer, delivering it to Bletchley Park in the back of a post office truck in January, 1944. Code and Cypher School staff dubbed Flowers's electronic symphony "Colossus" on account of its size, and no fewer than ten would be running by the end of the war, with "Mk II" versions incorporating up to 2,400 tubes. Most computer scientists agree Colossus did not yet meet the full spec of a Turing Machine, because it was built for "special" rather than "general" purpose calculations. Neither was there internal memory to support "stored program" architecture. But Flowers's creation continues to be a source of fascination to computer scientists, and in 2009 a University of San Francisco professor named Benjamin Wells established that the cluster of ten machines running at Bletchley by war's end, working in parallel, would have been "Turing complete." Among the first computer's many indispensable contributions to the defeat of Nazism was confirmation that Hitler had been fooled into expecting the D-Day landings at Calais rather than Normandy. Supreme Allied Commander Eisenhower received the relevant intercept from Bletchley and ordered invasion for the next day. World War II *might* still have been won without Colossus, but historians of the period agree it would have lasted years longer and cost countless more lives.

🐞 🐞 🐞

John Von Neumann went to England in late April, 1943, a month after Alan Turing returned from his six-month stint in the United

States, while work was underway on Colossus. There is no explicit record of his having visited Flowers's team or Turing at Bletchley, despite a hint of the latter in a letter to his IAS colleague Oswald Veblen, which also admits meeting Max Newman: official histories reveal only a trip to His Majesty's Nautical Almanac Office in Bath to view the non-classified computing program there. Given the intense secrecy surrounding Bletchley and related operations, it is not far-fetched to imagine the Bath trip as a front. What we do know is that von Neumann came home enthused by computing and joined the Manhattan Project straight away.

American computing did not start with von Neumann. The difficulty of modeling optimal trajectories for every combination of shell and gun deployed by the US Army, to create "range tables" for gunners, had already incited work on a first "general purpose" computer at the Moore School of Electrical Engineering, University of Pennsylvania. Up to that point range tables were compiled by stables of human computers, but their already painstaking job was getting harder and slower as weapons technology advanced, until the system was seizing up. With men absorbed into the forces and math one of the only technical disciplines open to women at university level, nearly all two hundred "computers" at the Moore School were female. When a team was needed to work out how to set up the new mechanical computer for each calculation it was tasked to do—this assumed to be a menial job by the exclusively male hardware engineers—the human computer pool seemed the natural place to look.

The ENIAC (Electronic Numerical Integrator and Computer) was built around a staggering 17,468 vacuum tube switches and weighed a cool thirty tons. Urban legend has lights dimming across Philadelphia whenever it was switched on. Like Colossus, ENIAC initially had no internal memory, but unlike Colossus, it was "Turing complete," meaning it could in principle be programmed to calculate anything that was calculable, if only anyone knew how. In the mid-1940s there wasn't even a word for what we now understand as "programming."

Six young women were selected: Kay McNulty, Betty Jean Jennings, Betty Snyder, Marlyn Wescoff, Fran Bilas and Ruth Lichterman. Their chief mentor and point of contact with the designers and engineers was twenty-four-year-old Adele Goldstine, whose Air Force mathematician husband, Herman, was ENIAC project administrator. Adele had a masters in mathematics from the University of Michigan and was working on a doctorate when called with her husband to Pennsylvania. She taught math within the Moore School and was the only woman to work with ENIAC's hardware team, eventually cementing her place in the annals of computing with authorship of the machine's three-hundred-page manual. In an oral history prepared by the computer scientist W. Barkley Fritz for the Institute of Electrical and Electronic Engineers (IEEE) in 1996, one of the ENIAC programmers, Betty Jean Jennings, an adventurous twenty-year-old from rural Missouri when she arrived at the Moore School, recalls encountering the magnetic Goldstine for the first time in a class on "inverse interpolation":

> *I'll never forget the first time I saw Adele. She ambled into class with a cigarette dangling from the corner of her mouth, she walked over to a table, threw one leg over its corner, and began to lecture in her slightly cleaned up Brooklyn accent. I knew I was a long way from Maryville, MO, where women had to sneak down to the greenhouse to grab a smoke.*

Men and women alike seem to have revered the Bacall-like Goldstine. An accomplished violinist, the "whip-smart and exuberant" mathematician brought music, warmth, style and intellectual rigor to the Moore School and later became a respected collaborator with John von Neumann at the IAS (where his wife Klára also did important programming work). According to Goldstine's daughter, Madlen, the two families remained close for generations, while von Neumann's daughter, the economist Marina von Neumann Whitman, says of Goldstine's involvement in her father's postwar work:

> *Adele spent quite a bit of time at our Princeton house . . . it was clear that [my father] regarded her as a significant participant in the project. She sometimes carried the infant Mady in her arms, and as I was a high-school girl, gave me my first example of a young woman enthusiastically embracing new motherhood and carrying on with her professional career at the same time. She gave me confidence to do the same when my time came.*

When the ENIAC women were introduced to their machine in 1945, there was no manual and wouldn't be for another year. They were flying blind. Most engineers had worked on their own specialist area of the machine: only the two designers, J. Presper Eckert and John Mauchly, and perhaps the Goldstines, understood the circuitry's mazy modular logic—and even they only knew it in theory. Tutelage for the young ENIAC programmers consisted of being handed an armful of wiring diagrams and told to figure it out. Nothing about ENIAC was easy. Changing a program could take weeks, as individual tasks were mapped to the circuitry of the machine using plugboard cabling and almost four thousand ten-way switches. In effect, the programmers were handcrafting algorithms, creating a new Turing Machine for each new job, doing what the magic of the modern "stack" does for you and me.

Through the novelty and importance of their work the ENIAC women grew close, like a band, the last programmers in town. And also the first. According to Jennings, they "worked together, lived together, ate together," and shared everything.

> *I mean, they were wonderful. We were crazy about each other. First of all there were two Jews, one Catholic, one Quaker, and I was a member of the Church of Christ, and so none of us had ever been around people of other religions like that. So we spent a lot of time talking about religion as well as about the ENIAC. And of course they took me to Baltimore so I could eat lobster, and Ruth took me to New York, and Marilyn took me to Washington, because I hadn't seen these places. So I really had a wonderful time with them.*

Soon the ENIAC women knew their machine better than anyone.

"Occasionally, the six of us programmers all got together to discuss how we thought the machine worked. If this sounds haphazard, it was," Jennings explains,

> *The biggest advantage of learning the ENIAC from the diagrams was that we began to understand what it could and what it could not do. As a result we could diagnose troubles almost down to the individual vacuum tube. Since we knew both the application and the machine, we learned to diagnose troubles as well as, if not better than, the engineers.*

According to a later addition to the team, Homé Reitwiesner, the women sometimes received unsolicited help from outside sources, not least the university cleaners, "who would on occasion reconnect one or more of the cables into any convenient open position" after knocking them out with their mops.

More helpful was the eagerness of engineers to explain the parts of the machine they had responsibility for, and the encouragement of John Mauchly, one of ENIAC's two designers, to view his system's potential through the widest possible lens. ENIAC was a machine you could walk around and through like a stand of trees; could touch and see in a very direct way. The women grew to know and love it as no one else could. W. Barkley Fritz notes that "little attention seems to have been given by the designers as to how the programmers were to do the job of using ENIAC to solve real problems." But the ENIAC women learned quickly, and it is not unreasonable to wonder if, by war's end, they understood computing better than anyone bar Turing.

Contemporaneously with Grace Hopper at Harvard, the ENIAC programmers are credited with the first use of *subroutines* (or *functions*); of *nesting* (placing one function inside another); of the *loops* that define computing to this day. There were no programming *languages* yet, meaning that where we can set up a Python function simply by typing the keyword *def*, for "define a function," so mobilizing screeds of hidden code to help us, the women were wrangling logic and circuitry in rawest form. Subroutines evolved out of fear that some ballistic trajectories would be too complex for the circuitry to simulate, according to Jennings. Her colleague Kay McNulty, who would go on to marry John Mauchly, solved the problem with a flash of insight into the computer's power. "Oh, I know, I know," she said in excitement one day. "We can use a master programmer to repeat code." *Perform this calculation ten times, or until the combined product reaches one hundred, or until no more numbers are left to perform on . . .* thus was the "while loop" born.

From that day, Jennings continues:

We began to think about how we could have subroutines, nested subroutines and all that stuff. It was very practical in terms of doing this trajectory problem, because the idea of not having to repeat a whole program, you could just repeat pieces of it and set up the master programmer to do this . . . once you've learned that, you learn how to design your program in modules. Modularizing and developing subroutines were really crucial in learning how to program.

These were some of the few women spared Rosie the Riveter's fate as female workers were pushed back into the home after the war. When recognizable programming languages began to develop in the 1950s, providing shortcuts for common procedures, the legendary Grace Hopper and ENIAC alum Betty Snyder (by now Betty Snyder Holberton) would work together to blaze trails in that field, too. Jennings recalls telling her friend Holberton that Grace Hopper had called her "the greatest programmer she ever met," to receive in reply: "Well, you know Grace—if you can do anything better than she can, you have to be the best in the world!" None of the ENIAC programmers knew their contribution to computing would be forgotten for half a century and remain little-known even now.

The first serious test problem run on ENIAC, in December 1945, modeled a thermonuclear explosion for von Neumann's team at Los Alamos. According to project administrator Herman Goldstine, the program took a month to run and involved a million data punch cards. Yet von Neumann knew ENIAC would not be powerful or flexible enough to facilitate the hydrogen bomb research he led as the Cold War loomed (very much against the wishes of most IAS peers and their spouses, one of whom recalls writing STOP THE BOMB in the dust on the windshield of his Cadillac). Colleagues found him sketching ideas for the kind of machine he would need, spidering some of what his friend Stan Ulam calls "the first flow-diagram coding" on blackboards.

The computing narrative grows murky around this time, enough for me to spend several days trying to piece together a plausible view of what happened next. John von Neumann is best known to computerists for lending his name to the architecture, "von Neumann architecture," that defines modern machines. He joined the ENIAC group in 1944 and was intimately involved in discussions about a "stored program" successor, which, consistent with computerers' collective genius for unlovable acronyms, would become known as "EDVAC" (Electronic Discrete Variable Computer). Before ENIAC had been completed, in June 1945, von Neumann wrote his *First Draft of a Report on the EDVAC*, setting out with his customary clarity the logic of the proposed computer—and most every computer to follow. Von Neumann hand-wrote his hundred-page draft on the train between Los Alamos and the IAS, then mailed it to Herman Goldstine in Pennsylvania. We know he didn't intend it for publication at that point, because he repeatedly refers to sections that were never written.

Goldstine had the notes typed up and distributed internally to people involved in the project. So exciting was the captured vision that it began to be sent around the world, still with no attributions or contextual notes. The glamour of von Neumann's name led many to assume *First Draft* had sprung from him alone.

Von Neumann very clearly did not believe—or wish others to believe—that the principles of digital computing were his. Members of his team at the IAS were in no doubt that von Neumann regarded Alan Turing as conceptualizer of this new field, with *On Computable Numbers* given as unconditional required reading for all who worked with him. In a letter to the computer scientist and historian Brian Randell, the Los Alamos physicist Stanley Frankel wrote:

> *I know that in or about 1943 or '44 von Neumann was well aware of the fundamental importance of Turing's paper of 1936 . . . von Neumann introduced me to that paper and at his urging I studied it with care. Many people have acclaimed von Neumann as the "father of the computer" (in a modern sense of the term) but I am sure that he would never have made that mistake himself. He might well be called the midwife, perhaps, but he firmly emphasized to me, and to others I am sure, that the fundamental conception is owing to Turing, in so far as not anticipated by Babbage.*

In fact—as we shall see—Turing's conception was more like Ada Lovelace's than Babbage's or von Neumann's. At any rate, he seems to have been unconcerned with attribution and was happy to cite *First Draft of a Report on the EDVAC* in his own detailed proposal for a new stored-program computer, presented to the UK National Physical Laboratory in February 1946. Von Neumann's name would henceforth be synonymous with classical computing, to the extent that Kurt Vonnegut's satirical first novel, *Player Piano*, published in 1952 and set in a dystopian postwar America where labor has been automated away by the all-powerful EPICAC supercomputer, turns on a rebellion masterminded by a character given the mischievous name "Professor von Neumann."

First Draft did cause tension between von Neuman and Herman Goldstine on one side and ENIAC designers J. Presper Eckert and John Mauchly on the other. Whether deliberate or not, failure to credit others' work is a cardinal sin within the academy, so umbrage was bound to ensue. But there was a complicating factor in Eckert and Mauchly's emergent wish to patent the prime features of the EDVAC for themselves. In Goldstine's 1972 personal history of the period, *The Computer: From Pascal to von Neumann*, the ENIAC administrator and mathematician wrote:

> *When the discussions leading up to von Neumann's work on the EDVAC had taken place it had been against a background of complete mutual openness and desire to produce the best possible ideas. Later it turned out that Eckert and Mauchly viewed themselves as the inventors or discoverers of all the ideas and concepts underlying the EDVAC. This view was strenuously opposed by von Neumann and me.*

Von Neumann considered the important features of the EDVAC to derive from Turing, but neither he nor the Englishman is on record as ever having contemplated patenting their ideas. None of which implies bad faith on the part of Eckert or Mauchly: when they began work on ENIAC and first considered a "stored program" successor, they appear not to have been aware of Turing's work—and could not have known about Colossus (as von Neumann almost certainly did) because its existence remained secret until the 1970s. All the same, in distributing von Neumann's paper, Goldstine had placed EDVAC's workings in the public domain and made patenting impossible. Relations soured, and Eckert and Mauchly left the Moore School to found the first US computer company, The Eckert-Mauchly Computer Corporation

(EMCC), at which ENIAC programmers Jean Jennings Bartik and Betty Snyder Holberton would be prominent employees. EMCC delivered the first commercial computer, the UNIVAC, to the United States Census Bureau in 1951. But for Herman Goldstine's precipitate action in distributing von Neumann's report, Eckert and Mauchly might have owned the classical computing paradigm and the US industry developed in a very different way.

Here we come to the crux. In the hundred pages of *First Draft*, von Neumann fleshed out the concepts Turing described in *On Computable Numbers* in a way that was explicit and clearly actionable. Central to the architecture he described was internal memory: the ability to perform a calculation (say, adding two numbers) and then temporarily store the result for later use (say, to add to a third number).

In an ideal world this moment would be marked either by fanfare or a bloodcurdling scream, because with the advent of memory we're at base camp; *on the metal*, in the realm of machine code. Classical computing as we understand it starts here, with a stupidly simple piece of circuitry called a *flip-flop* or *latch*. Fashioned out of four relay switches that make or break contacts within a circuit, a latch can be in one of two states at any moment, giving it the innate ability to retain a single binary digit (*bit*) of information—a one or a zero—until it is next needed. There is nothing unusual in the electronic latch's ability to function in this way: a dropped piece of toast, which can fall buttered side up or (more usually) buttered side down, has the same inherent ability to retain a zero (say, down) or a one (up). In other words, if we left our toast buttered side up on a counter, to represent one, then left the building and returned a week later, it would still signify a one to us. If we left two pieces next to each other, we would have the ability to create a number with two bits.

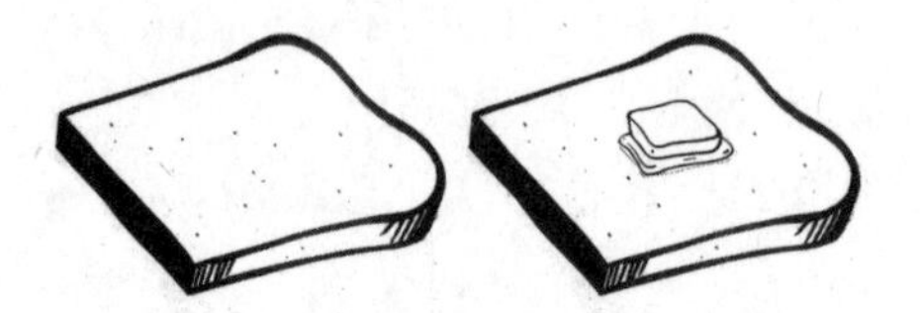

Fig. 01: the binary number "01" given in toast.

How would we read this binary number? For anyone rusty on number systems, they're not hard. The decimal system we use for most things is called *base ten* in the world of math. As we know, it utilizes the symbols 0, 1, 2, 3, 4, 5, 6, 7, 8, 9. Reading a decimal number from the right, its places indicate tens to the power of zero (i.e., *ones*) in the first place and tens to the power of one (i.e., *tens*) in the second place. Thereafter we find tens to the power of two, meaning 10 × 10 (hundreds); to the power of three, or 10 × 10 × 10 (thousands); to the power of four, or 10 × 10 × 10 × 10 (ten-thousands), and so on, as below:

Fig 10: the number of holes in Blackburn, Lancashire

Binary or *base two* numbers work the same way, but use only two symbols—zeros and ones. Reading from the right again, the first place indicates ones, with the second representing twos, the third twos to the power of two (in other words, *fours*); the fourth twos to the power of three (*eights*); the fifth twos to the power of four (*sixteens*), etc. A binary *byte* equivalent to the decimal number 42 looks like this:

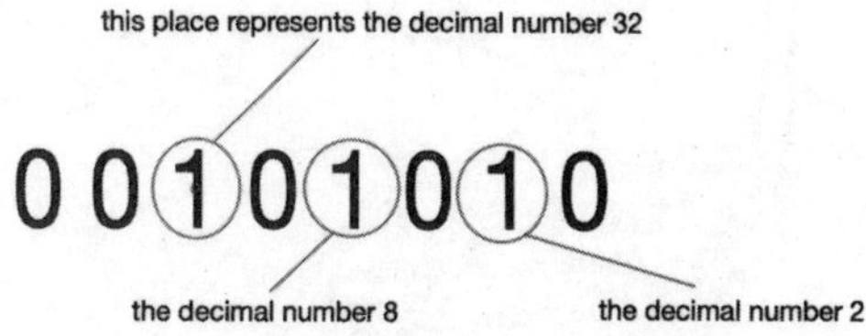

Fig 11: the decimal number "42" in binary form . . . 32+8+2=42

Or, given in toast:

Fig. 100: "42" expressed in binary form through the medium of toast

So why aren't computers made of toast, we might ask, affording them a deliciousness they don't currently have? The truth is that computers could be made of toast, they would just be larger and on the slow side. In principle, given enough space, a digital computer could be built from anything rearrangeable and with two possible states: sticks, utility bills, burritos, lotus petals, yaks, wheels of cheese, formation helicopter teams, Elvis impersonators, tiny coders . . . the possibilities are as finite or infinite as the amount of time we're prepared to wait for an output.

Fig. 101: the binary number 00101010 (42 decimal) stored by means of Elvis impersonators

Fig 110: computer built with tiny coders as memory latches

We don't need to be rocket scientists to guess that electronic latches will be more convenient units of computer memory than yaks or tiny coders. But whatever we use, the state of the machine at any given moment is the sum of the states of these latches, each with their own tiny meaning, a zero or one, much as a beach is the totality of its grains of sand. Unlike grains of sand, however, each latch in a stored memory computer has a numeric address, allowing its bit to be accessed and used or overwritten with each cycle of calculation. When combined, four bits is a *nyble* and eight bits is a *byte*; 1,024 bytes is a kilobyte; 1,024KB a *megabyte*; 1,024MB a *gigabyte*; and 1,024GB a *terabyte*, or close to a trillion bytes. If we want to get really intense, a *petabyte* is around a quadrillion bytes, and an *exabyte* roughly a quintillion—one followed by eighteen zeros. Anything more is a *whatever-obyte*, because no human mind can begin to embrace such magnitude.

So memory is a lot of small, simple circuits served by a central processing unit (CPU), much as houses are served by a postal service. But memory is only useful because its state can change easily by flipping latches, unlike Colossus or ENIAC, which had no internal memory and therefore no possibility of software. Those machines had to be reprogrammed for each new task by physically rearranging their circuits with cables, switches and dials. So it is that the beating heart of the stored-program computer Turing saw (and von Neumann elaborated) is the *clock*. The clock in a computer is an oscillator that actually does beat like a heart, moving between on and off, completing and breaking a circuit very rapidly. Von Neumann imagined this effect could be achieved using vacuum tubes or a rapidly vibrating crystal, with each change from off to on and back constituting a single *cycle*. As explained earlier, in a stored-program computer, each cycle does one minuscule piece of work. *Load* the byte you find at *this* address into the *accumulator*; *add* the byte you find at *that* address to the accumulator; *store* the product of these two bytes in *this other* address.

Cycles are organized like musical triplets: the first cycle is reserved for instruction—do *this*—while the second cycle contains the first half of a memory address and a third carries the second half of that address. And so it goes, round and round, unchanging forever if desired. So: lots of numbers flying around. Or more accurately, lots of tiny physical changes taking place at a microscopic level, to which we allocate meaning. This is machine code. Ask how these numbers are generated, using collections of switches von Neumann postulated as Boolean "E-elements" but which we now call *logic gates*, and the mind really starts to reel. See the end of this chapter for an explanation of this frankly bizarre process.

It would be reasonable to wonder why we're here, trying to understand the way a computer speaks not to us, but to itself. Most programmers will say we don't need to understand the internal logic of the machine. And they're right. But there is a point, and that point is for us to grasp the truth of how *surreal* the whole "von Neumann" classical computing mechanism is; what a weird, roundabout way to accomplish things, with billions of microscopic switches shuffling invisible galaxies of pretend ones and zeros around a slab of silicon, in the belief that these can be made to map a world full of mess and emotion and analog processes that took eons to evolve by entirely different means.

The real shock comes with the realization that computing is built entirely of metaphor and works only as a gargantuan metaphor—perhaps the most elaborate ever conceived. For instance, there is nothing remotely like a "loop" inside a computer. There are only numbers that combine to perform functions our minds are capable of *imagining* as loops—of "picturing" in this way, if you will. Nothing on a chip and almost nothing in a programming language is real in the sense of having more than a tenuous, figurative relationship to what we call it. Only the consequences are real. In this sense programming *is* like magic, casting a symbolic spell across the world. How not to be awed by such a confection? As an expression of human creativity, the microcosmos seems to me unparalleled.

And let's look at the improbability of the clock: speeds are measured in *Hertz* (Hz), meaning "cycles per second." The Intel chip in my two-year-old laptop clocks at a mundane-sounding 2.3 gigahertz (GHz). That's right, *yawn*. Except what this means, I now understand, is that it performs 2,300,000,000 cycles per *second*—What! *How?*—and as with the spinning of a celestial body like the moon, you have to ask: Why wouldn't it just stop one day?

We haven't even reached the you-must-be-joking stuff yet. My laptop contains 32 gigabytes (GB) of memory, which translates to an ability to hold 256,000,000,000 binary digits (each a one or zero) in 32,000,000,000 numeric memory addresses, assuming each holds one 8-bit *byte* (as is usual). That's 256 *billion* bits to be generated, slung around, retrieved on the whim of a human who probably has no idea what actually happens in this freshly instantiated, rapidly evolving cosmos governed by its own rules and conception of time.

Some of the logical maneuvers undertaken to make this work are spectacular. How long would a binary number equivalent to address 31,999,999,999 in my laptop's memory be? It would be 35 places long. Which is why the machine code used internally to address memory and convey instructions is not usually expressed as binary, but as *hexadecimal.* Also known to mathematicians as base sixteen, *hex* utilizes the symbols 1-2-3-4-5-6-7-8-9-A-B-C-D-E-F, with A–F representing the decimal numbers 10–15. In hex terms, my outsize binary memory address becomes 773593FFF. This explains why memory is usually given in multiples of sixteen (16, 32, 64, 128, 256, 512, 1024 . . .) rather than in the decimal or binary increments one might expect.

Hex numbers are generated using collections of switches known as 16-bit counters and are written with a lowercase *h* suffix. Within the *instruction cycle* of a particular microchip, the number 10h might signify "load," after which 20h might accord to the instruction "add," 11h "store," and FFh "halt." Different chip manufacturers use different machine code *instruction sets* with different *opcodes* (operation codes), but experts agree the most momentous instructions are a cluster of variations on the theme of "jump." Once you can jump from one place in a program to another upon the fulfillment of a specified condition, then you can loop—and as Charles Petzold notes in his book *Code: The Hidden Language of Computer Hardware and Software*, "Controlled repetition or *looping* is what separates computers from calculators."

Petzold's book is the most accessible I've found on the phantasmagoria of the machine, explaining in detail the *half adders, full adders, accumulators, multipliers, three-input AND gates, level-triggered* and *edge-triggered D-type flip-flops*, *2-line-to-4-line decoders*, *8-line-to-one-line data selectors*—and on and on—with improbable lucidity. Yet, from the intense month I spent reading *Code* at the start of this weird trip, an entire week went to three pages explaining how hexadecimal addresses emerge from these generic arrays of binary switches. To characterize the logic of this process as tortuous doesn't begin to do it justice: my margin notes read like a lost scene from *The Silence of the Lambs*. But it *is* jaw-dropping. Up close, the microcosmos is that rarest thing: a human-made artifact which, on being demystified, seems not less but more like wizardry. Nature also contains fantastical systems. At a molecular level, the processes of metabolism or of DNA transcription are mind-blowing. The difference is that these mechanisms are part of a universe thirteen billion years in the making, within which we took four billion years to evolve and adapt—not a few decades of hopeful human fumbling. Did we expect this to go smoothly?

Wartime innovators including the German Konrad Zuse went on to develop the first stored-program machines, many following the blueprint laid down in von Neumann's *First Draft*. Vacuum tubes defined the hardware until transistors took over in the mid-1950s, at which point computers began the relentless process of shrinking in size and growing in power to the point where they fit in our pockets; on our wrists; literally in our hearts; anywhere at all. This doesn't mean that trailblazers all saw the future of computing in the same way. Where von Neumann tended to be concerned with building powerful functionality into his machines' hardware, Alan Turing envisioned the physical machine as a flexible, minimally specified vessel for revolving galaxies of code, which is why he was perpetually thinking ahead to what he insisted should be called "mechanical intelligence" and predicted programming would be a fascinating pursuit, where von Neumann saw dull clerical toil.

In a phone conversation, the retired embedded software expert Jack Ganssle, who writes a reflective and entertaining blog called *The Embedded Muse*, tells me about a thought experiment he conducted, where he imagined an iPhone built from ENIAC technology. He starts by setting NASA's Vehicle Assembly Building (VAB), one of the largest buildings on Earth by volume, as a unit of size. From here, he estimates that the ENIAC iPhone would occupy space equivalent to 470 VABs and weigh as much as 2,500 Nimitz-class aircraft carriers—a useful deterrent to use while driving. On the other hand, Ganssle points out, given that our phone's cost would equal the GDP of the entire world, finding anyone to call might be tricky anyway.

The influential software engineer and instructor Robert C. Martin, known as "Uncle Bob" on YouTube, riffed on Ganssle's calculations in a 2016 talk on "The Future of Programming." But his purpose was not to impress us with the strides made in computing hardware in a single lifetime, it was to point out the extent to which software hasn't kept up. Hardware has transformed out of recognition at least twice since the 1950s, Martin notes, while there have been *no* radical advances in software writing technology. If a C programmer could hop in a TARDIS and go back to 1974, they would still be able to practice. With a little refinement the reverse would also be true.

"Code is assignment statements, if statements and while loops," he says. "It has been that way and it will be that way: that's what code is."

This is because the bedrock of our computing remains the stark machine language developed to suit the electronic components available in the 1930s and at the outset of World War II. Everything on top of that is what the British call a "bodge" and programmers fold into the noun-verb *kludge*. Programmable computing machines were needed quickly and could be cobbled together using electrical switches, which were well understood and could be manufactured and pressed into service *fast*. As the shadow of fascism fell across the world, who had time to muse, *Is this the only or best way we could do computing?* The people who launched computing and programming remain heroes of the highest order: they got a hard job done. At the same time, von Neumann wasn't the only wartime pioneer to recognize the drawbacks of the misnamed "von Neumann" approach, and he assumed we would find better ways to compute going forward. The tragedy of Alan Turing's loss at the age of forty-one in 1953, closely followed by von Neumann three years later at just fifty-three, is that the deeper philosophical and practical questions these visionaries had been asking at the time of their deaths fell away for the rest of the twentieth century, leaving a lot to catch up on in the twenty-first.

Julian Bigelow worked as chief engineer on von Neumann's computing Meisterwerk at the IAS, completed in 1952 and usually known as the IAS machine. Von Neumann published everything and patented nothing, so his machine was much studied and widely cloned (including by IBM). Yet both men well understood the IAS machine's structural weirdness, according to Bigelow. In *Theories of Memory*, his short history of the IAS project, he observed: "The modern high-speed computer, impressive as its performance is from the point of view of absolute accomplishment, is from the point of view of getting the available logical equipment adequately engaged in the computation, very inefficient indeed." Most parts could perform constantly at high speed, but "are interconnected in such a way that on the average almost all of them are waiting for one (or a very few of their number) to act."

Bigelow was describing what John Backus, the illustrious lead developer of the first popular high-level programming language, Fortran, referred to in a 1977 Turing Award Lecture as the "von Neumann bottleneck." Simplified, this "bottleneck" refers to the characteristic of classical computing architecture whereby instructions and memory addresses must take turns in the *duty cycle*, like kids on a zipline (Backus used the metaphor of a "tube"). He suggested some possible solutions in terms of language development, which Edsger Dijkstra, true to form, summarily blew out of the water. What Dijkstra didn't do in any serious way was quarrel with Backus's oft-quoted characterization of the machine's workings as "primitive." Note that by *store*, Backus means the memory, while *word* refers to a single unit of data:

> *Surely there must be a less primitive way of making big changes in the store than by pushing vast numbers of words back and forth through the von Neumann bottleneck. Not only is this tube a literal bottleneck for the data traffic of a problem, but, more importantly, it is an intellectual bottleneck that has kept us tied to word-at-a-time thinking instead of encouraging us to think in terms of the larger conceptual units of the task at hand. Thus programming is basically planning and detailing the enormous traffic of words through the von Neumann bottleneck, and much of that traffic concerns not significant data itself, but where to find it.*

All those addresses flying around the system contribute nothing to the computing of a problem. In *Theories of Memory* Julian Bigelow revealed that he and von Neumann were already looking past classical architecture, to "the possibility of causing various elementary pieces of information situated in the cells of a large array (say, of memory) to enter into a computation process without explicitly generating a coordinate address in 'machine-space' for selecting them out of the array."

In other words, the mechanism favored by biology. Nature doesn't work with specific and precisely defined addresses, but with templates, things that fit together or connect, allowing all elements to be sorting at once. Computers are fast but linear, while biology is slow but distributed—and may provide a clue to the long-term future of computing. As George Dyson notes in *Turing's Cathedral*:

> *Bigelow questioned the persistence of the von Neumann architecture and challenged the central dogma of digital computing: that without programmers, computers cannot compute . . . The last thing either Bigelow or von Neumann would have expected was that long after vacuum tubes and cathode-ray tubes disappeared, digital computer architecture would persist largely unchanged from 1946.*

Nonetheless, von Neumann could be unnerved by the machines he was building. In that unpublished memoir of her husband, Klára reports him coming home from Los Alamos in an agitated state one day, skipping meals ("nothing he could have done would have had me more worried") and sleeping for an unusually long time, before waking up in the middle of the night, perseverating over the computing power he was unleashing.

"What we are creating now is a monster whose influence is going to change history, provided there is any history left," he said, according to his wife, who continues, "While speculating about the details of future technical possibilities, he gradually got himself into such a dither that I finally suggested a couple of sleeping pills and a very strong drink to bring him back to the present and make him relax a little about his own predictions of inevitable doom."

In the desperate months after the German Navy added a fourth rotor to their Enigma machine in 1942, a force of six thousand mostly Canadian troops attempted a raid on the German-occupied French port of Dieppe for reasons few could fathom at the time. More than one historian now claims the raid to have been a diversion, overseen by the future James Bond author Ian Fleming, then attached to naval intelligence, with the true goal of capturing one of the new four-rotor naval Enigma codebooks. The risky mission was unsuccessful, with a loss of life Canadians still mourn. Fleming, waiting offshore aboard the destroyer HMS *Fernie*, went home bereft. Yet the capture of an Enigma codebook would have been deemed worth the loss. We have to assume that, but for the secret computing army at Bletchley; the Bombe; Colossus, high-stakes Allied ventures like Dieppe would have been far more common.

Fleming liaised directly with Bletchley Park on other schemes to capture an Enigma codebook or machine, and the cancellation of one such mission is said to have left Alan Turing and his specialist team in despair. The author was fascinated enough by the Bletchley operation to leave highly illegal hints of its existence in at least two of his Bond novels, *From Russia with Love* (where Bond is lured into a trap on the promise of a stolen *Spektor* encryption machine) and *You Only Live Twice*. According to Bletchley researchers, Fleming's description of the ideal leader for his Enigma commando snatch team amounts to a point-for-point description of Bond.

In the early 1990s, Bletchley Park, having eluded Nazi bombers throughout World War II, came within an inch of being destroyed by the Luftwaffe's spiritual heirs, property developers. The late computer scientist Tony Sale launched a successful campaign to save and renovate the site for the world, simultaneously leading a team of enthusiasts in a decade-long labor of love to reconstruct a working version of Colossus, now on display at the adjacent National Museum of Computing. The latter task was never going to be easy: after the war, the Bletchley Park Colossi were dismantled, with their parts scrubbed of clues as to their provenance and recycled. A pair of final machines went to the Government Communications Headquarters (GCHQ), where they remained in covert service for another fifteen years, before finally meeting the fate of their forebears. Heart breaking, Tommy Flowers acted on post-victory orders to destroy all records pertaining to his machine, later lamenting, "That was a terrible mistake. . . . I took all the drawings and the plans and all the information about Colossus on paper and put it in the boiler fire. And saw it burn."

The Official Secrets Act fell across Bletchley and everything that had happened there. For the next thirty years, the public would have no knowledge of the measureless impact of Alan Turing or Tommy Flowers, or the tireless and uncatalogued army of women codebreakers and programmers who contributed as much to the defeat of fascism as had Battle of Britain pilots. Having made a pledge not to speak of their exploits, many refused to talk even when they legally could, seeking no thanks or credit beyond personal knowledge of what they had combined to achieve.

Photos at the site show a large and vibrant community drawn from every walk of life and including male and female linguists, classicists, mathematicians, scientists, engineers, writers and poets. The Government Code and Cypher School's enlightened commander, Alastair Denniston, knew his people needed relief from the mental strain of their work, so there were lively amateur dramatic and choral societies, along with Scottish dancing, fencing, chess, bridge and skating on the lake in winter. There were romances and marriages of all kinds. Alfred Dilly Knox, who made the first breaks into the Enigma code with Turing in the Stableyard, is reported to have been the lover of John Maynard Keynes while they were together at Eton. You get the impression that the contingency of war made people somehow freer. Turing would be persecuted for his homosexuality in the 1950s, a cause for the greatest and most eternal national shame, but there is no evidence of any such thing happening at Bletchley.

Why do I get choked up as I gaze at the faces in the pictures, smiling in improvised costumes for a production of *A Midsummer Night's Dream*, or lost in concentration as they race to translate an intercept that might be meaningless or save ten thousand lives, or tend the miraculous Colossus like sparrows in a hedgerow? There is a personal connection, sure. I was born in New York, but my parents were children in England during the war, forced to evacuate with their schools from London and Hastings on the south coast (both hometowns of Turing's, by coincidence)—an experience that changed their lives in lasting ways, and with theirs, mine.

But that's not it. For a time I think my brimming emotion stems from being in the presence of such awe-inspiring creativity, in service to something so obviously *right* that it can only be a cause for wonder—and maybe even hope. But that's not it either. Only when I find myself standing before a memorial to three Polish mathematicians do I start to get a handle on the odd mix of lightness and dark augury I feel here. What I learn is this:

Six years before the war, the Polish Cypher Bureau trio of Marian Rejewski, Henryk Zygalski and Jerzy Różycki broke Enigma by hand and even built a machine to speed the process called the "Bomba." When fear grew that their country would be overrun, they smuggled their work to the French and British. The Bombe machine Turing and Gordon Welchman built to automate the breaking of Enigma had not sprung from nowhere: it was informed by a device the Polish Cipher Bureau had built previously. Moreover, the perforated paper sheets used by the human codebreakers in Hut 6 were called "Zygalski sheets" after their inventor. Outside of Bletchley Park, Polish pilots died in droves for the RAF; Polish soldiers fell in the failed raid on Dieppe. And not just Polish: citizens from dozens of countries across Europe and the world, whether directly affected by Nazi expansionism or not, gave everything they had. For the sake of others.

I stand staring at the memorial and think of the Polish cryptographers, of Turing and so many others who died with almost no one knowing what they had done, nor apparently feeling anyone needed to know . . . who regarded what they did as its own reward, with no reference to a datastream no one can touch or feel or ascribe meaning to outside of itself but that somehow became the prime arbiter of value and truth, in whose presence humans are reduced to nodes in a network—Dijkstra's objection to Engelbart, I suddenly understand. Ten thousand people worked here, were prepared to work so selflessly and sacrifice so much in these drab and uncomfortable conditions for the sake of something intangible and yet still real, without ever demanding, "What's in it for me? What do *I* get?"

The desolate thought occurs that such a collective effort of will is hard to imagine outside these gates now. And yet, that our alienation from each other and ourselves may owe something to the revolution that began in innocence here. In the first half of this book we made the difficult journey down to the metal, which raised lots of questions. Now I want to turn and travel back up the stack, into the world in search of answers.

🐞 🐞 🐞

The final section of this chapter is an optional extra for anyone curious to go deeper into the microcosmos. There is no reason at all to try and internalize what follows—the aim is to further illustrate how idiosyncratic our way of computing is. Those already convinced—or happy to take my word for it—should feel no guilt about executing a swift conditional jump to the next chapter, which will be about the connection between code and adorable puppies.

We recall Boole's revelation that the whole of reality can be represented symbolically using just a few algebraic tools, namely:

(a) ones and zeros, to respectively indicate *True* or *False*; *on* or *off*; *everything* and *nothing*;
(b) the basic operators *and*, *or*, *not*;
(c) variables to which values may be assigned (e.g., y = dogs named "Kevin")

We also know that between 1932 and 1938, electrical engineers including Claude Shannon had the insight that Boolean logic could empower electrical circuits to do more than convey electricity; that they could do math, process data, maybe one day *think*. A common way to illustrate this is with a simple circuit consisting of wires, a pair of switches arranged in series, and a lightbulb. When the circuit is closed, electricity flows freely and the bulb lights. If one or both switches are opened, however, the circuit is broken and the bulb goes off. A diagram of this scenario could look as follows, with the pair of switches marked (a) and (b) respectively:

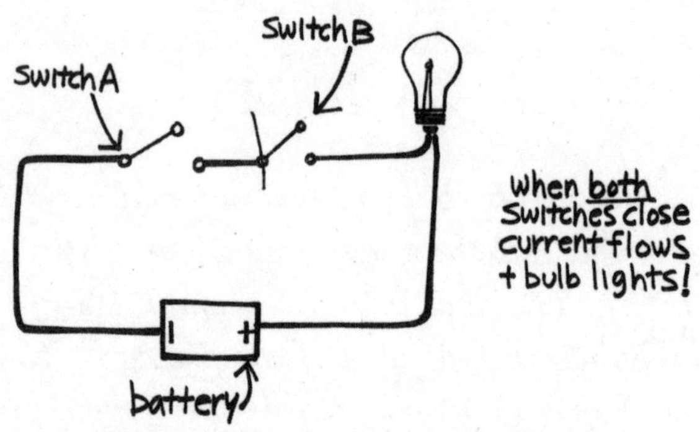

Fig. 111: circuit diagram for an AND gate

Now assume that both switches are attached to an electrical input of their own, and that when current flows, the switches close. When the flow to both switches is on, the lightbulb will be on, but if either is off, the bulb remains dark. From here we can impose the symbolism of a Boolean function, because if we take *on* to represent "True" and *off* to mean "False," we can see that if both inputs are True, the circuit will evaluate to True, but if either is False, the output will be False. Equally, we could impose a numeric value, with "on" representing *one* and "off" indicating *zero*. In this case, only two ones will produce an output of one, with anything else returning zero. This, the most basic calculation unit of a classical computer, is called an AND gate. We can produce a *truth table* to describe its output.

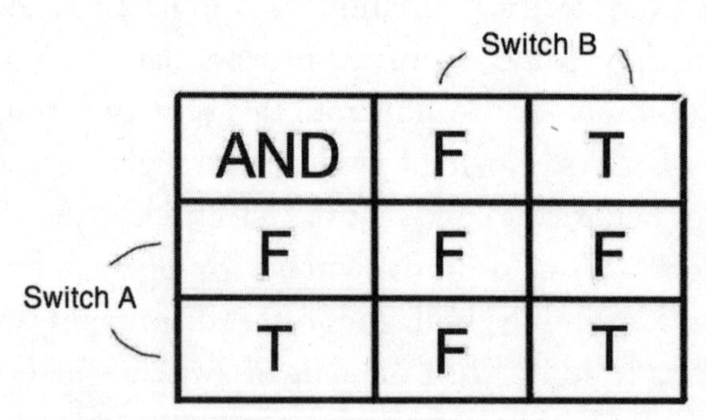

AND	F	T
F	F	F
T	F	T

Fig. 1000: An AND gate truth table. Only if both inputs are positive/true is a positive output produced

Now consider a circuit wired in parallel, like this:

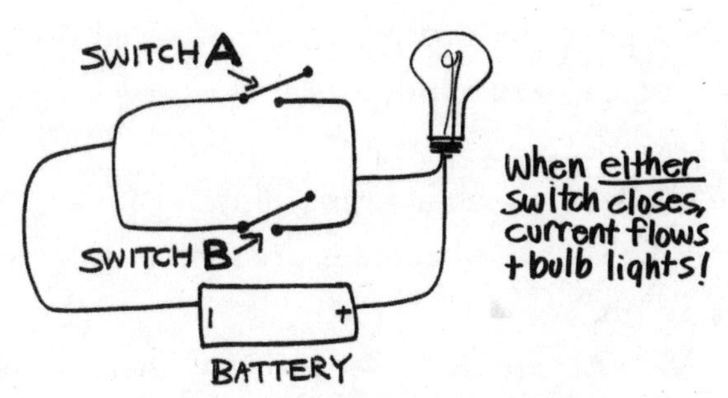

Fig. 1001: circuit diagram for an OR gate

We see that the output of this gate will be positive if *either* input is positive (or both are.) Which is to say $1 + 1 = 1$, but $1 + 0$ also equals 1. Here we have an OR gate, with a truth table that looks like this:

OR	F	T
F	F	T
T	T	T

Fig. 1011: An OR gate truth table.

It's easy to see that AND and OR gates could have multiple inputs. An AND gate with sixteen inputs would require all sixteen to be True in order to produce an output of True, for instance. A four-input OR gate would confirm that at least one of the inputs was True. The possibilities are endless.

What we have here is switching theory. Play with the logic and it becomes possible to produce circuits that do almost anything. Other logic gates include

1. The *inverter*, which inverts an input so that True becomes False, False becomes True, zero becomes one. And so on.
2. The *NAND gate* (for "not and"), which evaluates to True only if both inputs are *False*.
3. The *NOR gate* (for "not or"), which outputs True only if both inputs are False or both are True.
4. The *XOR gate*, which evaluates to True only if both inputs are different: if they are the same, be it True or False, the output will be False.

Computerists use symbols for these gates so they don't have to draw a circuit each time they refer to one. The symbols for AND, OR, NAND and NOR gates respectively (with *Q* representing the output) are:

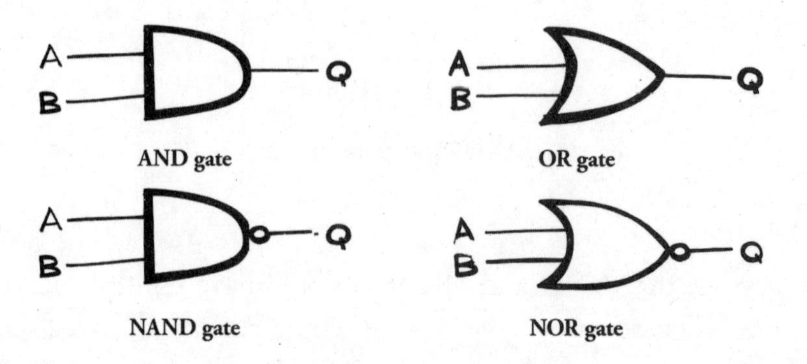

Note that in using these symbols, we no longer need to think about the gate's internal logic—we can take for granted what it does and are free to start chaining it to other gates to perform more complex operations. In hiding the detail of how the job gets done, we have just created the world's smallest black box, containing the most primal unit of computing.

Hilarity Ensues

"Forty-two!" yelled Loonquawl. "Is that all you've got to show for seven and a half million years' work?"

"I checked it very thoroughly," said the computer, "and that quite definitely is the answer. I think the problem, to be quite honest with you, is that you've never actually known what the question is."

—Douglas Adams, *The Hitchhiker's Guide to the Galaxy*

I still remember the best lecture I ever attended. It was part of a joint series offered by the English and philosophy departments in my first semester of college and, given the subject was Sartre's *Being and Nothingness*, should have been the dullest event in Europe that night. But it wasn't. The lecturer, Thomas Baldwin, had a deceptively simple style: he would write a proposition on the blackboard and gaze at it for a moment, like a medium beckoning a spirit. Then he would turn and smile and start to explain.

Baldwin paced the room—but slowly. On occasion he would stop altogether, appearing lost, a moment in which all the world's logic seemed at stake, before somehow refinding his path to a second thrilling proposition. At one point he stood with his forehead in his hand for so long we almost called for a medic. He was so engaged, so present, that you could almost *feel* the motion of his mind—and through his, your own. To this day, if I'm feeling blue, I think back to Baldwin's explication of the logical transition from existential anguish to nausea and invariably I feel better.

Baldwin's talk came to mind a few years ago when I listened to a BBC radio debate about lecturing standards at British universities. I then had two children at college, both of whom found lectures frustrating, so the contention of UK education minister Jo Johnson that quality in this area was "highly variable" came as no surprise to me. What's more, during sample lectures on open day visits, I had the same experience of being bored to tears by things I should have enjoyed. So when my daughter reported an exception to this rule, I'd seen enough to know what my first question would be.

"Did the lecturer use PowerPoint?"

"Hm. No, he just spoke," she said.

PowerPoint was then so ubiquitous that Lotte hadn't made the connection. But the lectures I attended had left me in no doubt that Microsoft's wildly successful presentation program, as typically then used, was not just inimical to, but destructive of deep thought: could have been scientifically designed to put the eagerest mind to sleep. The more I inquired into why this might be, the more I recognized a case study in the way even relatively passive code, in its seductive cloak of convenience, skews our relationship to the world in ways we can be slow to see.

How so? PowerPoint enthusiasts claim it emboldens nervous speakers and forces everyone to present information in an ordered way. Both contentions are true. But the price of these advantages is that in a conventional presentation the speaker dominates an audience. Where the space around and between points on a blackboard is alive with possibility, the equivalent space on slideshow screen is dead. Bullet points enforce a rigidly hierarchical authority that has not necessarily been earned: one accepts them *in toto* or not at all. And by the time any faulty logic is identified, the screen has been replaced by a new one as the speaker breezes on, safe in the knowledge that yet another waits in the wings. With everyone focused on screens, no one—least of all the speaker—is likely to be internalizing the argument in a way that tests its strength. It is possible to draw upon and customize PowerPoint "slides" in real time, but through most of the program's existence hardly anyone has.

So, a few bored students: how serious is this? If the problem ended there, the answer would be *not very*. But it doesn't—and a glance at PowerPoint's origins helps explain why.

The genesis story runs like this: from the late 1950s corporations began to realize that, rather than going to the trouble of developing new products they hoped would meet a need, they could use marketeers to create the perception of need and then develop products to meet it (a shift dramatized in the TV series *Mad Men* and Sam Mendes's lauded National Theatre production of *The Lehman Trilogy*, where it is attributed to the eponymous banking brothers themselves). To do this, different departments had to be able to speak to each other, to sell ideas internally. So while there had always been meetings, now there were meetings *about* meetings and—hey presto!—the modern world was born.

The presentational precursor to PowerPoint was the overhead projector, which is why its screens are still called "slides." The program owes most to Whitfield Diffie, one of the Time Lords of online cryptography, but it was quickly snapped up by Microsoft. Written largely in C++, PowerPoints's coding and marketing roots are intrinsic to its cognitive style, which is relentlessly linear and encouraging of short, affirmative, jargonesque assertions—arguments that always appear resolved and untroubled by shades of gray, uncoincidentally redolent of function definitions more than satisfying discourse.

It's also no coincidence that the two most famous PowerPoint presentations are (1) the one offered to NASA managers by engineers, explaining with unarguable illogic why damaged tiles on the doomed space shuttle *Columbia* were nothing to fret about; and (2) General Colin Powell's equally fuzzy pitch for war with Iraq. It goes without saying that blaming PowerPoint for Iraq would be like blaming Darwin for Donald Trump, but the program made scrutiny of the case harder. Not for nothing did then Brigadier General H. R. McMaster subsequently liken the proliferation of PowerPoint presentations in the military to an "internal threat," saying, "It's dangerous because it can create the illusion of understanding and the illusion of control. Some problems are not bullet-izable."

Perhaps even worse in our twenty-first-century circumstances is a charge leveled by the French writer Franck Frommer. Because PowerPoint can only present propositions and arguments as equations, he says, they appear to have no owner: in another curious echo of the underlying medium, algorithmic binary code, no one need feel responsible for them. Post-banking crises, we know how dangerous this perception can be. Many speakers now follow Steve Jobs's ever-adroit early example by restricting PowerPoint use to pictures, sounds and video. At PyCon I watched several wildly entertaining presentations delivered in this way, one using cartoons that must have taken ages and much thought to prepare. PowerPoint had made these successful talks possible, but it hadn't made them easier.

The point is this. Even simple, innocuous-seeming software packages can and do direct us in complicated ways, ways that reflect the motion of the medium through which they are expressed. Moreover, where their benefits tend to be foreseen, downsides often appear out of nowhere; seem to rain unannounced from the sky.

One of the lessons I'm learning is how hard it is to do code well.

A month after my brain scan, I am treated to a life-affirming illustration of the degree to which ostensibly simple tasks can be complex when it comes to code. My latest idea for a project is to design a new website using the open-source Python web framework Django. To prepare, I've decided to attend the annual DjangoCon US conference in San Diego, which is where I find an Australian developer named Russell Keith-Magee giving a talk on the subject of time.

Keith-Magee, a Django core developer and former president of the Django Software Foundation, has one of those minds that crackles with intelligence. He begins by explaining his particular interest in horology, the science of timekeeping, with an observation that humans have always struggled to fold the celestial movements governing earthly seasons into a usable calendar. A fundamental difficulty is the fact that the solar year, meaning the interval from summer solstice to summer solstice, is not exactly 365 and 1/4 days, as assumed by the "Julian" calendar Julius Caesar introduced in 45 BC. A leap day every four years therefore meant that by the sixteenth century the solar year was ten days out of step with the calendar, which created problems with the date of Easter and would have wreaked havoc on spring break down the line.

Astronomer and mathematician advisors to Pope Gregory XIII aimed to solve the Julian problems by introducing the Gregorian calendar in 1582. Now there would be a leap day every four years, unless the year was divisible by one hundred, when a leap day would *not* occur . . . except that if the year was also divisible by four hundred, it would. Okay. October only had twenty days in the Catholic diaspora that year, but Easter was now calculable, with spring break and the great state of Florida safeguarded for future generations. The problem for programmers is that the non-Catholic world followed suit very much at leisure, with forward-thinking Sweden coming aboard in 1700 but getting the math wrong and being forced to add an extra-extra day to February 1712—leading to a one-off instance of the suckiest known birthday, February 30. Meanwhile, few will be surprised that the British Empire held out until 1752, or that Turkey prevaricated until 1926. Keith-Magee notes that this week marks the one-hundredth anniversary of Red October, a.k.a. the October Revolution in Russia—an event that, perhaps fittingly given Stalin's ultimate ascendency, happened in *November*, because only after the revolution did the Julian calendar follow the Czar and Eastern Orthodoxy out a Kremlin window. The broad point being that the number of days in a year, or in February, or even in October, can vary depending on (a) the year being considered, (b) where the programmer is and (c) wherever an end user might be.

"Now, this is an amusing story, but I want to make a point," Keith-Magee continues. "Like a lot of problems in computer science, time is something that seems relatively simple, like 'What is your name?' and 'What is your sex?' The number of days in a year can seem trivially easy to implement, but in practice you need to understand a lot of human history if you're going to implement a robust solution."

Computerists refer to the unusual but predictable problems at the margins of normality as *edge cases*. As problems, edge cases are usually explicable and solvable, but they can require disproportionate consideration. When programmers joke, "Great! We're 90 percent done . . . only 90 percent to go," this is part of what they mean.

"You just have to be aware that there are edge cases and pay attention to them," Keith-Magee continues. "The problems we see on a daily basis in computing dates and time handling is because people either don't understand the complexities of problems, or they don't care, or they don't communicate the limitations of the solution they've used, or they're willing to make those limitations someone else's problem in the future."

An example. The so-called "Y2K bug" at the end of the last century stoked fear that all computer systems would crash or malfunction as clocks hit midnight on January 1, 2000. The panic had its origin in two rudimentary oversights. First, early programmers had misunderstood the Gregorian calendar's exception to the exception that leap days would not occur in years divisible by 100, failing to note that 2,000 was divisible by 400 and therefore *would* get an extra day. More serious was the fact that, at a time when computer memory was still expensive, programmers decided to represent dates using two digits instead of four, meaning the machines had no way to distinguish the year 2000 from the year 1900. Both errors appear to have emanated from the 1970s: what began as a space-saving optimization devolved to "a major engineering headache" in the last years of the century.

Lesson learned—or not. A similar convenience hack caused AOL's servers to fail in 2006, according to Keith-Magee, and a potentially more serious reprise of the Y2K problem is due at 3:14 on the morning of January 19, 2038 (sometimes called Y2K38):

> *But here's the thing. All computer systems have limitations. . . . The problem with AOL's server wasn't that they used a hack to keep dates from expiring: the problem with Y2K wasn't that the system used two characters to store a year. The problem was that the techniques that were used set a hard deadline for the end of life of that code. And the end of life wasn't understood or institutionally indicated. As a result, when the clock ran out, hilarity ensued.*

Python, like other languages, has libraries full of packages (or *modules*) to help with these issues, and Keith-Magee spends time discussing their merits and how best to deploy them. In doing so, the true depth of the challenge starts to become clear. Dates come in many different formats, varying by country, culture and time zone. Time is often given as UTC (Coordinated Universal Time) or GMT (Greenwich Mean Time) plus or minus a number of hours, with an additional hour where appropriate for daylight savings time. But Keith-Magee points out that knowing what time-zone a user is in says nothing about what language they speak; how they expect to see their dates formatted; what time it is locally *right now*.

Standardization came with the railways. Until 1840, Bristol in the west of England kept a clock timed to its own solar noon, ten minutes after that of London. The Great Western Railway was thus the first to adopt coordinated time, set to that of the Royal Observatory at Greenwich, although GMT didn't become the legal default until 1880. A similar process took place in the United States from the 1880s, with time zone borders often running through train stations. The gusty burghers of Detroit kept their own discrete local clock until 1900 and only settled on Eastern Standard Time in 1915.

Time zones are an artifice; a convention we observe for our own sanity, pitting the grand sweep of the heavens against our puny mathematical whims. As such, they can change on a dime. In 2018 North Korea gave five days' notice of hopping time zones to synchronize with South Korea. Keith-Magee notes that in his own backyard, Darwin (in the Northern Territory of Australia) and Adelaide (in the south) both set their clocks to UTC plus nine and a half hours—except that the latter institutes daylight savings time in summer, while the former does not. The town of Broken Hill is in New South Wales, but observes South Australian time, while retaining New South Wales's daylight savings transition dates: these usually coincide, but not always. New South Wales observes daylight savings, where neighboring Queensland does not. Lord Howe Island does observe the change, but offsets by only thirty minutes. Other continental anomalies abound.

"And we haven't even left Australia yet," the developer laughs.

Where daylight savings is observed, clocks go forward in spring. That night, deducting ten minutes from 3:05 A.M. in the United States will render not 2:55 A.M., but *1:55*, making simple mathematical functions (deduct ten minutes from *x* time) useless to the programmer. But it gets trickier still. Just as there are leap days, there are also leap *seconds* to bring the calendar back into alignment with the solar day—an adjustment that has caused major software crashes in the past. To mitigate the risk from leap seconds, the New York Stock Exchange halts trading for sixty-one minutes around a transition. Even within the Anglophone world, cultural perspectives matter. When an American company announces a software release for summer, meaning a summer month in the Northern Hemisphere, "everyone south of the equator rolls their eyes." If we are going to write effective code, we need to take all of the above information into account *and* know the location and specific circumstances of the person who is using our code, at which point "all the same problems happen all over again."

There is a broader lesson still. When using automated tools to help navigate this or other programming minefields, Keith-Magee argues, a coder must remain alert to both what they are doing and what society is doing around them. Most code libraries and modules are engaged in a balancing act between flexibility and absolute accuracy, so will occasionally be off. No amount of fancy logic will tell you whether 081018 is October 8, October 18, or August 10. The deciding factor can only be the totality of circumstance *in the world*.

"So, if you use these [date-time] libraries, be aware that they are not magic wands. They make assumptions and those assumptions have consequences. And that is not a bad thing. All code makes assumptions: you just need to be aware of what assumptions your code is making and validate those assumptions as reasonable," he concludes.

I leave the lecture thinking, thank God for the reflective coders—though, in truth, most of the coders I've moved among so far match this description to one degree or another. It increasingly seems to me that the mind of a good programmer needs to be as panoptic and sensitive as a historian's or a philosopher's or a novelist's. Likening the production of software to "engineering" might be doing them and us a disservice. In fact, the reasons we do call it "engineering" turn out to have significance, as I am about to find out.

On the second day of conference I inadvertently feed a wrong address into Google Maps. It's a bright morning and the pale eggshell sky mirrors my cast of mind as I process a *New York Times* report headlined "Soldiers in Facebook's War on Fake News Are Feeling Overrun." An *LA Times* slung onto the back seat suggests e-scooter startups have become a bubble. Thus distracted, I am even more inclined than normal to follow Google Maps wherever it tells me to go.

Only after leaving the highway do I gasp and realize my mistake on being confronted with a scene I had no way of expecting: a grid of dusty streets and crumbling houses, prowled by a twitching forest of figures huddled in layer upon layer of rags until they are two or three times the size of whatever human frame might remain underneath. In a dreadful echo of post-apocalyptic cinematic convention, the figures' feet drag as if magnetized to the pocked asphalt, with skin so tight and eyes so occluded that it's hard to imagine anything resembling life behind them. For a moment I feel paralyzed, unsure how to act or feel. I grope for empathy, but find only a bolt of dread, as if my wrong turn has been into some kind of dismal presentiment. Shocked, I drive till I find a place where it feels safe enough to stop and reengage Google, then get out fast, feeling confused as the day I was born, unable to process such abjection, such *horror* in the cradle of affluence and entitlement that is California.

Ten minutes later, still shaken, I am strolling through reception at the Mission Valley Marriott, while Django programmers stare at screens in the pleasant beige courtyard, sometimes alone and other times in groups, recommending talks or exchanging tricks, happy to be in each other's company—as am I, feeling guilty at my own relief. What none of us know as we survey this urban-bucolic scene is that, behind it, Marriott management are grappling a data breach compromising the security of five hundred million customers. Lucky as we are, the safe ground under our feet seems to be iterating away, too.

Django is interesting. It's called a "framework" rather than a language because it's a collection of complimentary code modules, written in Python, that fulfill many of the common and boring tasks of creating websites, allowing developers to concentrate on the creative stuff. It was named for the great Romani-Belgian Hot Club de France guitarist Django Reinhardt because one of its two originators is a Ukrainian American journalist and guitarist named Adrian Holovaty, a Reinhardt fan who moved to Amsterdam and believes in "journalism by computer programming." The other originator, Simon Willison, a British CS grad with a similarly adventurous resume, met Holovaty while interning at the *Lawrence Journal-World* newspaper in Kansas from 2003–2004. That a web framework used by Instagram, Mozilla, LinkedIn, *The Guardian* and NASA could have been developed at a local Kansas newspaper seems hard to believe just fifteen years later. As DjangoCon began, the *Journal-World* stood as a rare local paper hanging on through the infowars.

I'm not the only one to notice that the highly international DjangoCon crowd is more balanced in terms of gender and race/ethnicity than other code gatherings, with a higher proportion of communicative, socially comfortable people. An entertaining roster of talks reflects this prismatic energy, taking in engineering ethics; programming with attention deficit disorder; reasons not to dismiss JavaScript as horrible; the coding scene in Africa; and using Python to explore gender bias in children's books, not to mention the inevitable machine learning. On the surface this openness looks like—and in itself *is*—something to celebrate, because enormous effort goes into making the Django community welcoming and inclusive. Only when I get a chance to sit down with a veteran coder named Diane Chen, active in the local San Diego chapter of DjangoGirls, do I start to see how much more complicated the truth is.

The broad strokes of Chen's coding origin story are everywhere at DjangoCon. Arriving at college as a noncommittal biology major in the 1970s, she met her first computer program while analyzing lab results.

"And I just thought it was so cool. You'd give raw numbers to the software and get this interesting output. So, I took a programming class, in Fortran, and I had so much fun in that class that I took another computer class—and it was all over! I just loved it," she says.

She changed her major and took a bachelor's, then master's, in computer science. To her, the process of computing always seemed creative, like solving a puzzle with a fixed set of tools. As with most programmers of her generation, Chen swung through the orbits of many languages, including assembly, Perl, Java and C++, at one time working on a project to convert ten million lines of superannuated Fortran code into C ("I'm not kidding . . . that was an *adventure*"). She thinks the Python and Django communities are more diverse than most because they welcome beginners—noting that the same is true of the culturally similar Ruby, whose popularity owes much to its own successful web framework, Ruby on Rails.

Chen's perspective on code culture is meaningful, because she was forced to withdraw for a decade through 2013 to care for a mother with Alzheimer's, then return to the field she loves as a woman in her fifties—a coding demographic equivalent to ruminative shock jocks or shy politicians. Refinding her place was hard, so she completed C++ and Java certificates at the University of California San Diego and started going to Python meetups. Even so, when she got a job at a local tech behemoth, it was working in Perl, a language primed to eclipse all others in the late 1990s, before collapsing like a tower of Jenga blocks. There were some good programmers at the new gig, she allows, but the mode of the language seemed to encourage pride in the production of "obfuscatory" code incorporating long chains of impervious statements separated by semicolons, creating obscure ways to do simple things. C coders of the 1980s had often frustrated her in the same way, mistaking convolution for sophistication.

"That was always one of my pet peeves," she says. "I couldn't get anybody interested in best practices or standards, or anything like that. Why would you want to write code no one else could understand or work with?"

Chen's Perl code had been in service of Android phones, so when her yearlong contract was up, she decided to study Android with a view to going in that direction. She continued going to Python meetups, where she had fun, made friends and even joined a robotics project. But when she went to the equivalent Android events, she found a very different atmosphere.

"Out of fifty or sixty people, maybe two were friendly. It wasn't that the rest were *un*friendly, just nobody made a point of being friendly. Everybody seemed anxious and crabby. And frustrated. The last one I went to turned out to be right before a Python one, so I'd just had this really unpleasant experience, followed by a nice one. And I thought 'What are you doing?' So I dropped everything else and said 'I'm just going to learn Python.'"

She bought the books and took the classes and now teaches Python, mostly to natives of other programming languages, work she enjoys. And yet Chen's most jarring revelation is an impression that, in the ten years she was away from her beloved trade, the situation for women got *worse*. Chen estimates a third of her colleagues to have been women when she started out—and when I check the figures, they bear her out. In 1984, women accounted for almost 40 percent of CS degrees, with Bureau of Labor Statistics figures from 1987 indicating a similar ratio in the programming workforce. But if the rate of decline between then and 2015 had continued, there would have been *no women programmers* by 2020. As it happened, the hemorrhage hit a floor at 5 percent in 2015, creeping back to a reported 7 percent as Chen and I talk, and falling back toward 5 percent by 2021. As noted this loss of female talent makes programming unique among professions, which have nearly all moved toward greater inclusivity.

"Back then, I worked on CAD/CAM [Computer Aided Design and Manufacturing] software that was used by General Motors, Pratt & Whitney and McDonnell Douglas," Chen explains. "So our user group consisted of mechanical engineers and aerospace engineers and automotive engineers that traditionally you would assume would be more prejudiced. But what I found—and the women all found—was that, although they might assume you didn't know anything to begin with, as soon as they saw you did, they were fine. And I mean, I didn't find any of the kind of prejudices that I felt in 2013, where the group I was in was 90 percent men. I was always the last to know about anything, and if I ever tried to say anything, I would just be interrupted and steamrollered, just talked right over. And if you complain, you get told you're being oversensitive, you know? That old 'bitch' thing . . . all that garbage."

She continues, "I think the difference is that even in the eighties and nineties, although some men might be surprised you were a woman, once they saw your ability and experience, they respected it. Whereas in 2013 it felt more like they were threatened and wanted to keep you out, would exclude you and set you up to fail. It was more of a power trip."

We know the first programmers were overwhelmingly female. Code lore from the early days acknowledges as much. Grace Hopper was the Data Processing Management Association's first "Man of the Year" in 1969—the word "person" having yet to be invented—while the much-garlanded MIT computing brain Margaret Hamilton ran a mixed-gender team writing software for

NASA's Apollo Guidance Computer.[1] An emergent literature has begun to unearth stories like those of the ENIAC women, which were little noted at the time and buried for decades thereafter. Meanwhile, historians like Nathan Ensmenger, Janet Abbate, Marie Hicks and others have begun to explain how the Great Purge happened—a baroque story whose legacy should alarm anyone reliant on software.

From what we know so far, the Purge occurred in two distinct waves, with roots in the 1960s and 1980s, respectively. Distilled from a disparate range of sources,[2] a summary narrative could run as follows:

By the late 1950s, corporate management teams and governments understood that hardware was junk without good software. Suddenly programmers became important, but there weren't enough to match computing's breakneck rise. The trouble was that few 1950s managers knew anything about code or coders. Most practitioners of this abstruse new art came to it haphazardly, so who even knew where to find them? Or how to assess their skills? Something had to be done but no one knew what—fifties shorthand for *time to call a psychologist.*

Up until then IBM recruitment ads had been broadly framed. A *New Yorker* "Talk of the Town" piece from 1959 cited one ad as targeting people who enjoyed "musical composition and arrangement," or playing chess or bridge, or who had lively imaginations and curiosity. Now the psychologists swatted this humanistic approach aside. Based on a small study of professionals with an average age of twenty-five and fifteen months' programming experience, the IBM Programming Aptitude Test (PAT)

---

1 And saved the first moon landing as a direct result of being forced to take her daughter to work, when the girl exposed a software problem while playing in the lunar lander flight simulator. (See Notes & Sources.)

2 For sources and further elucidation, see Notes & Sources.

they devised relied heavily on multiple choice questions involving math, some predicated on formal mathematical training. A key assumption was that coding minds are born and not made, in the way of mathematicians or musicians—a view lent spurious credence by a flawed later report claiming good programmers to be *twenty-five times* more productive than bad ones. "10X" lore starts here.

Soon after PAT appeared, a Bureau of Labor report found skepticism among employers as to its real-world relevance, not least because higher-level languages like Fortran and COBOL had removed most of the math from coding. The test became an industry standard for the next two decades nonetheless. By 1967, with panic emerging over a supposed "software crisis," more than 700,000 people were invited to take PAT on the off chance they were blessed with one of those rare coding minds. At one point, the entire population of Sing Sing prison in New York was offered the test, while an article in *Cosmopolitan* invited readers to become "computer girls" in an industry by now "overrun with males" (which was read at the time as "likely husbands"), quoting no less an authority than Admiral Grace Hopper in assurance that programming was "just like planning a dinner." Even *Cosmo* couldn't help but notice that PAT's emphasis on formal math skills weighed against women, favoring the upper- and middle-class white men most likely to have had formal training. The anti-woman turn was exacerbated from 1965, when the Association for Computing Machinery (ACM) made membership conditional on a four-year degree, causing female membership to collapse from 40 to 10 percent within a few years. By the end of the seventies, quality research had exposed PAT performance as no better a gauge of coding potential than eye color, shoe size or knowledge of the works of J. R. R. Tolkien.

A further problem: laying a PAT test on anything with opposable thumbs and a pulse was irksome. But what was the alternative? More psychologists! Employers had long used personality tests to assess the suitability of applicants to particular jobs. Now a pair of academic psychologists named Dallis Perry and William Cannon used a series of papers with titles like "Vocational Interests of Computer Programmers" to compile a slate of personality traits they imagine distinctive of the trade. Except they didn't find enough traits, unearthing exactly one "striking characteristic" of the competent code jockey: disinterest in people. If you want to find born programmers and 10Xers, they concluded, go to where the people who don't like people are.

The amusing thing is that almost no one disagreed with the psychologists' assessment of programmers' personalities—even programmers. When IBM's John Backus led a team to develop Fortran in the mid-1950s, his aim was to give scientists, mathematicians and engineers a means to communicate directly with the machines, sparing interaction with a confederacy he personally coined the disparagement *coder* to describe. Even while flaying Perry and Cannon's problematic research at the 1968 ACM conference, the respected computing industry analyst Richard Brandon characterized the coding breed as "often egocentric, slightly neurotic" and "excessively independent," with a paranoid nod to schizophrenia. Brandon and others insisted that PAT favored mathematically inclined males by default and actively selected for such curmudgeonly personality types, so of course researchers would find them well represented among the fraternity. The sequence of distortions embodied in the PAT and personality tests in turn combined to create what the computer historian Nathan Ensmenger would call "a gender-biased feedback cycle that ultimately [selects] for programmers with stereotypical masculine characteristics."

Among the sternest contemporary critics of selection mechanisms favoring detached, asocial characters was the future

Psychology of Computer Programming author Gerald Weinberg, who identified a further reinforcing factor. Because coders of the pre-internet era tended to program alone, he suggested:

"The admiration of individual programmers cannot lead to an emulation of their work, but only to an affectation of their mannerisms . . . the same phenomenon we see in 'art colonies,' where everyone knows how to look like an artist, but few, if any, know how to paint like one."

Conference speeches, academic papers and articles from the second half of the sixties simmered with managerial elegies to the unmanageability of programmers who styled themselves artists and poets, disdained workplace dress norms and personal hygiene, and often had an antiwar countercultural bent, crystalized in the shameless disportment of beards.

The prescribed establishment remedy for coders' unruliness was professionalization. A 1968 NATO Conference on Software Engineering aimed to lay the groundwork for standardization and a new rigor. The Department of Defense set out to develop a programming language to meet its and the UK Ministry of Defence's needs, replacing the four-hundred-plus argots employed in its sprawling array of Cold War projects. That the military called its language *Ada* is at once sweet and weird, but from the early eighties to nineties it became the new field's rising star. By this time, however, the wooly stoners that IBM and the Department of Defense dreamed of bringing to heel were instead building an industry in their own image. Apple Computers' 1984 Superbowl ad for the first Macintosh, created by the *Alien* director Ridley Scott, invoked Orwell to present IBM as Big Brother. Touché.

The year 1968 saw a second ambiguous waypoint on the drive to "professionalism." Volume one of the Stanford academic mathematician Donald Knuth's *The Art of Computer Programming* series, published that year, provided an intellectual foundation for the study of what would become computer science. By claiming a lineage to the Persian mathematician al-Khwarizmi (Latinized to *Algoritmi*), Knuth posited his new science as the study of algorithms, more like math than electrical engineering. Volume one was called *Fundamental Algorithms* and inspired the ACM to design its *Curriculum '68*, which was heavy on theory and light on practical programming skills. Objective study of algorithms as the fundamental unit of computing, on par with gravity in physics or elements in chemistry, was a new idea. The word "algorithm" didn't appear in *Webster's New World Dictionary* until 1957.

Debate about the real-world utility of *Curriculum '68* and computer science as an academic discipline began immediately—and continued for decades to come. But these innovations did confer an aura of rigor seen as essential to professional recognition. They also presented a roadblock to young women and other disadvantaged groups at a time when math and science (and university education in general) was overwhelmingly white and male-dominated. Dispute over the proper relationship between programming and math would never end. And it could be worse. Where the displacement of women programmers in the United States was an implicit effect of the fight for professional status, a product of "laziness, ambiguity, and traditional male privilege, leading to the establishment of a highly masculine programming subculture," according to one later account, the UK purge constituted explicit government policy. Decades later the American historian Marie Hicks argues that there was a deliberate expulsion of women from programming in Britain, and the resultant leach of experience was a significant factor in the country's surrender of leadership in computing after World War II. There are parallels between Britain then and the United States now, she warns.

But wait! By the mid-1980s, computer science was established as the stock route to a programming career. And women's participation had recovered to the extent that parity with men in the near future seemed possible, even likely. Instead, the number of women in the field took a nosedive.

Hardly anyone noticed the hemorrhage, and most didn't care. An outlier who did was Allan Fisher of the School of Computer Science at Carnegie Mellon University. Alarmed that female undergraduate enrollment in his department had fallen to just 8 percent by 1995, with many women dropping out along the way, he engaged social scientist Jane Margolis to help discover why. What Margolis found was this: the advent in the early 1980s of relatively cheap home computers like the Commodore 64, Atari 800XL, BBC Micro and Sinclair ZX Spectrum gave rise to an entirely new genus of student—young men who had spent their teens playing electronic games designed by men like them. And through these machines they learned to hack. Now young women turned up for degrees to find themselves already lagging behind peers who arrived with the swagger of old hands and had little patience for less advanced classmates. Margolis found that family computers often resided in male siblings' bedrooms, with fathers consciously or unconsciously encouraging boys' interest at their daughters' expense.

Margolis and Fisher designed a program of remedial measures (including extra tutelage early in the program), which combined to increase female enrollment to a reported 42 percent in just five years, with a far higher proportion staying the course to graduation. One of Margolis's more intriguing assertions was that men and women at Carnegie Mellon appeared to approach their practice differently. While the men in her study group loved to hack, most women drew motivation and meaning from "the purpose that computing was going to be used for" (i.e., its social and practical applications).

Despite the success of changes at Carnegie Mellon, women's involvement in programming overall continued its collapse toward the 5 percent professional nadir of 2015.

It's a warm night in the open air. Djangonauts mill about and chat as the bar speakers fizz English eighties synthpop, the Stax soul of the computer age, made on machines more resembling Colossus than the software suites on MacBooks today. Hearing these sounds, I'm struck by (a) the unforeseen speed at which our technology has evolved since those last days of the analog age and (b) how naively optimistic the music sounds now, despite having been made at a bleak recessionary time for the nation. Also that (c) the most interesting electronic music of that time, including first wave New York hip-hop, was made with technology on the brink of obsolescence, on machines supplanted by newer ones, at which point the older machines became cheap and accessible. Technologists have a saying that "intelligence moves to the edge of the network," and the same may be true for creativity in other forms. That said, little of enduring worth gets made at the cutting edge.

I'm sitting for drinks with a developer named Anna Kiefer, trying to glean how gender imbalance plays out for a young woman at the coding coalface. Just shy of thirty and originally from D.C., Anna's path to code chimes with most I hear at DjangoCon.

Raised by a lawyer and a teacher, she studied journalism and political science at NYU but graduated to find quality journalism in freefall after the appropriation of ad revenue by Big Tech. Wanting to do something positive with her life, she joined a nonprofit and completed fieldwork in the Ecuadorian rainforest, where she saw the difference good software could make to a cause. A later renewable energy project required the development of a website.

"And that's where I got a little bit of exposure to front-end web development and web design," she beams. "And then from there I started learning on my own, and really enjoyed it."

Two years ago, she moved to San Francisco to enroll in an intensive twelve-week bootcamp, then set to the fraught business of applying for jobs in the code rush market of the Bay Area, eventually landing a spot as one of three engineers at a small startup. Even so, from where she sits the industry has entered a galling new phase for women. Revelations of deep misogyny at Uber and other tech firms over the past year were capped by the notorious "Google memo," sent by senior engineer James Damore to everyone at the company and claiming women have evolved to be inferior at tasks like programming. Kiefer fights hard not to betray anger at Damore's argument, knowing she walks the plank all objects of prejudice do: say nothing and the slurs stand; complain and you're uptight, difficult, no fun. Neither is hitching distortions of science to the cause of exclusion new. Victorian intellectuals denied women access to university based on pseudo-medical claims that it would disrupt menstruation and the reproductive system, needlessly endangering the species. Yet here we still are. These same nineteenth-century eminences claimed that exposing women to classical literature would lead to ungovernable licentiousness.

Damore's intervention will prove more interesting than I'm able to see right now, but as Kiefer and I talk I see that most of the disadvantage she experiences is subtler. The two male programmers she works with are nice and they all get on well: she seldom feels deliberately patronized or shunned. Instead, her outsider status manifests culturally, in an inability to participate in the things they do together, like going to comic conventions.

"So I don't know if I feel excluded per se," she reflects. "But I'm not *in*cluded."

With two years' startup experience behind her, Kiefer has been applying for jobs at the next level, at established companies, which is where the dearth of women in senior engineering roles starts to matter. She describes an ordeal that usually begins with a phone call from a recruiter. If that goes well, then one or two technical coding challenges will be presented remotely, as an engineer watches on a shared screen. Few programmers will have tried to code with someone watching—least of all a stranger, in a pressured situation—and it is unlikely ever to happen in real life. You get better at this, Kiefer says, but it never strikes her as an effective measure of ability.

"So much depends on whether they are good interviewers or not, or whether they're having a good day or a bad day. You can tell when you've got an experienced, more sensitive interviewer, because they'll do little things to help you feel more comfortable, explaining the process and telling you a little about themselves, engaging with you before you jump into the exercise."

Gentle, self-aware interviewers are far from a given. Some companies want two of these hour-long remote sessions before inviting the potential employee to an "on-site" that usually lasts five to seven hours. On-sites are rigorous and exhausting, Kiefer sighs, especially as the engineers assessing you are likely to be all men, from a demographic rightly or wrongly renowned for inhospitality to outsiders, not least women. Is there an "incel" or closet Damore in the room? Would they prefer a comics fan or gamer to go to conventions with? You don't know. Notoriously, examiners often ask candidates to solve coding problems by writing solutions on a whiteboard, something that, again, never happens in real life. Like the discredited Programming Aptitude Test of the sixties and seventies, these interviews appear to be selecting for qualities incidental to the work. Kiefer is far from alone in finding it hard to relax and shine in such an artificial situation, which could have been calculated to favor less empathic, socially sensitive people.

"I've noticed that I do things that are out of character [for me] when I feel uncomfortable in a room. My confidence goes down and I apologize a lot. Frequently, I notice after an interview that I've said 'sorry' thirty times throughout the day, which I don't usually do. And that probably makes me seem unconfident—which I'm not. There are many times, whether it's in the remote session or the 'on-site,' where I'll be stuck and sort of freeze up. And then, you know, twenty minutes later I'm by myself and can solve the problem easily."

She frowns. "A friend actually asked me, 'Do you think it's easier for you to get a job in this industry because you're a woman?' I had to take a deep breath and say, '*Are you effing kidding me!?*'"

A galaxy of research points to men (as a population) skewing overconfident and women underconfident in their abilities. It's easy to see why. A 2017 study of contributions to code projects on GitHub purports to show women's work being accepted more often than men's when gender is not specified, and less when it is. The study also turned up variations among languages, with acceptance rates significantly higher for women in Python, Ruby and C++, and lower in PHP and JavaScript. There is cause for cautious optimism, however. The nonprofit Mozilla Foundation, maker of the popular Firefox browser, responded to the GitHub study by developing a gender-blinding technique for their own contributors.

Some companies are tweaking the job interview format, too, trading whiteboard inquisition for take-home assignments.

"So you're not trying to code with somebody just staring at you who may or may not know how to interview effectively," Kiefer grimaces. "That seems much better."

Another important aspiration is to find a more sustainable work-life balance, she says, citing a problem endemic to Silicon Valley, where companies use the lure of free food and services to keep workers at the office and the culture applauds obsession. This is a passive form of discrimination against women, who may feel torn between careers and children, costing the industry yet more experience and breadth of perspective—not forgetting that fathers lose out here, too.

For all the trials, Kiefer doesn't for a moment regret her choice of career. Across two long conversations, our discussion covers every aspect of an art and profession she plainly loves. The constant learning inspires and excites her; she has the opportunity to use her skills to improve the world. A sense of empowerment and independence arises from this crazily expanding field. And of course there's the pay. She is now a *full-stack* programmer, meaning she codes both the *front-end* stuff we interact with when we open our laptops and the server-side *back-end* code that ties everything together and makes it work. Eventually she would like to move to the macro challenges of data and infrastructure architecture, or planning what teams of developers do.

"But, you know, I would still want to have a hand in the actual implementation. I can't see myself ever not coding."

At home I have a large and growing pile of books on the experience of women in Silicon Valley, plus an expanding stack on the relationship between race and technology, specifically race and code. One of these books, *Black Software: The Internet & Racial Justice, from the AfroNet to Black Lives Matter*, by Charlton D. McIlwain, is full of previously untold stories, perhaps the most astounding of which is this: In the late 1960s, at the request of presidential candidate Robert Kennedy, IBM made a sustained—and successful—effort to bring more women and minorities into computing at all levels. The company went so far as to set up a plant in the then-deprived Bedford-Stuyvesant neighborhood of Brooklyn, specifically to provide an on-ramp for Black people who may not have had one otherwise. Yet, within the coding cohort we know most about, in the United States, the percentage of Black coders relative to the overall population has been roughly equivalent to that of women in recent years. It goes without saying that for these groups, which combine to form about 64 percent of the US populace, these figures are bad (if unsurprising) news. Less remarked on is the disservice they do us all. Here's why.

Each year the coder's help site Stack Overflow carries out a survey of the profession, which is the best source we have for coding demography, and what emerges is a picture of staggering homogeneity within the profession. When DjangoCon is done, conference-goers will return to an industry in which 7 percent of coders are female and under 3 percent are Black. They will also find their peers' average age to be twenty-eight in the United States—and as low as twenty-two in India, with almost three-quarters having no dependents. Figures show that 80 percent of practitioners have less than ten years of professional experience and almost 60 percent fewer than five years. What's more, *close to a third have been on the job less than two years*. A veteran woman programmer describes having been at Google during a fire drill and being perturbed by the paucity of women and people of Central and South American heritage, by the near absence of Black people and shock of seeing "more dogs than people over forty." Map the above age and experience gradients onto aerospace, healthcare, architecture or newsgathering and ask: "Do I feel safe?" Now ask if what coders do is less important than, or even significantly different from, healthcare at this point in our evolution as a species.

Veteran coders tell us this wasn't always the case. In an address on "The Future of Programming," the previously mentioned engineer and teacher Uncle Bob Martin (born in 1952) asks his audience:

> *What happened? What's wrong with us that we are repelling half the people in the world? . . . In my first job there were 24 programmers, most of them in their 30s and 40s and half of them were women. But ten years later in 1980 my company had fifty programmers, all in their 20s and 30s, and only three were women.*

On one level, the twenty-first century age and experience skews are explicable. The industry is rapidly expanding, drawing younger people in. But that's not the whole story. Long workdays and 24/7 campus cultures will always weigh against people with children and/or full, mature lives, and this often affects women most of all. An amusing detail hidden in the 2018 Stack Overflow survey was its suggestion that half the population of professional coders spends less than one hour a day outside and more than nine hours at a screen (13 percent boast more than twelve hours). This situation may look advantageous to employers in the short term, because young single men who meet the MIT legend Joseph Weizenbaum's description of "compulsive programmers" probably output a lot of code and may ask fewer questions about the impact of that code. But such social and intellectual attenuation can only be at the expense of breadth of view, which is precisely what many tech companies need most as their products collide with the world. There *are* exceptions to the conventional pattern: at the collaboration platform Slack, energetic boss Stewart Butterfield, a former game designer and co-founder of Flickr who majored in philosophy, grew up on a Canadian commune and was originally named Dharma, made a point of insisting that under normal circumstances everyone had lunch together and left the office by 6:30 P.M. In January 2021, the firm was sold to the business software powerhouse Salesforce for close to thirty billion dollars, so his approach was not crazy.

We still haven't reached the heart of the problem, though. One of the most important stats gathered by Stack Overflow is tucked at the back of their report and easy to miss. It concerns not coders' education, but their parents', and is where we learn that 75 percent of programmers have at least one parent who went to university or equivalent, with 30 percent owning a post-graduate degree as well (for reference, one-third of US adults have undergraduate degrees as I write this). But that's not all. A 2017 analysis by a recruitment firm called HiringSolved pegs the top three feeder colleges to Silicon Valley as UC Berkeley, Stanford and Carnegie Mellon. Figures collated by the *New York Times* show a staggering 17 percent of Stanford undergraduates to be drawn from families in the top 1 percent of income distribution, with more than half from the top 10 percent. More shocking still is news that only one-fifth of Stanford students come from what researchers—with no apparent irony—dub the "bottom 60 percent" of the population in terms of income; of these, a meager 4 percent hail from the bottom fifth. A generous system of grants and scholarships at Stanford is not having the desired effect, unless the desired effect is to prevent the figures from looking even worse.

We also do well to recall that Stanford University, font of Silicon Valley, was founded by eugenicists and remained a hub for the propagation of that warped and self-serving worldview up to and beyond World War II—a war German Nazis (including Adolf Hitler) justified with reference to Californian eugenics theory and practice, with some citing California statutes in their defense at the Nuremberg trials. William Shockley, the Valley founding father credited with coinventing silicon-based transistors at Bell Labs, was a virulently racist and paranoid man who taught at Stanford and championed eugenics well into the 1970s, even running for the Senate on a eugenicist platform in 1982. Exclusion was very much a founding guest at tech's table.

What's more, the status quo at Carnegie Mellon is roughly the same as at Stanford. UC Berkeley, a state university, fares better, but not so much as you'd think. Thirty percent of students come from the "bottom 60 percent" at Berkeley.

Why is the coding population so strangled in terms of gender, race, age and experience? The clear answer is unequal access. At length I notice that the Black coders I get to know at DjangoCon, PyCon and elsewhere mostly hail from the eastern states, with few originating in California. When I share a coffee with Daniel, who I meet on the first day of conference, he tells me he works for a New York company called Broadway.com. In a team of thirteen developers, only two are CS grads. One came to New York to be a dancer, while Daniel himself studied electrical engineering—until he found programming.

"I know I don't fit the picture of what a developer looks like," he shrugs, "and I always expect I'll really stand out and be out of place at things like this. Then I come and see how much more diverse they are than I thought they'd be."

Does he think this is truer on the East Coast, or at least outside of Silicon Valley? "Probably, yeah," he says. Later, when over lunch I introduce Anna Kiefer to Cris, a dreadlocked Black coder from Philadelphia, noting that she recently moved from New York to the Bay Area, he grins, "Then you're an honorable exception, because most of the people who move from the East to the Bay Area are people I've been happy to see go, never the ones I don't want to lose." Everyone laughs, but I don't think he's joking. My suspicion is that the greater gender, race and age diversity at DjangoCon reflects not just a conscious policy of openness within the community, but the fact that front-end development (of the type Django facilitates) sits with academic research at the bottom of the status and pay stack. Current top slots go to engineering managers, site reliability engineers, data scientists and machine learning specialists.

The point I want to make is not about fairness. It's about jeopardy. A few years ago I saw a Hay Festival talk by Sarah Jayne Blakemore, professor of cognitive neuroscience at University College London, whose area of study is the adolescent brain. No one who has been a teenager, much less raised one, needs telling that adolescents can be both wonderful and vexatious. Common traits include moodiness, self-absorption and attenuated empathy. Painfully intense friendships and exaggerated risk-taking vie with equal drives to invention and originality, and to the brave pursuit of identity and meaning.

Professor Blakemore's groundbreaking research, elucidated in a book called *Inventing Ourselves: The Secret Life of the Teenage Brain*, turns up something important. Adolescent brains are constitutionally, *physically* different from the adult brains they will become. Young people can't help the changes that come over them at this time: they are in the throes of reconfiguration; of optimization, if you like. But here's the twist. Most brains do not complete the physical transition from adolescence to maturity before their mid—and sometimes late—twenties. Of special significance is underdevelopment of the prefrontal cortex, which is imperative to impulse control and understanding the consequences of actions, and has been postulated as a factor in the high incidence of young mass shooters in the United States. In this light, Facebook's "move fast and break things" ethos starts to make sense, as does founder Mark Zuckerberg's early claim that "Young people are just *smarter*," because Facebook was essentially conceived by adolescents—in fact, highly privileged adolescents of a type likely to comprise more than half of Silicon Valley's workforce. No wonder "move fast and break things" turned out to mean "do whatever you want and break everything." Why wouldn't it? Facebook was and is the business equivalent of an unusually dull spring break party.

Then there's the "voices in the room" issue. Examples abound of software compromised by the homogeneity of the teams developing it. Search online for "racist soap dispenser" and you find the "smart" restroom dispenser that used infrared light technology and worked for white but not brown hands, thanks to an absence of brown-skinned people in the room when it was being developed, while some early voice recognition and videoconferencing software struggled with vocal timbres typical of women. All of which might seem shruggable until you remember that software often works invisibly behind the scenes, approving loans, sifting job or college candidates, authorizing benefits or identifying suspects. Circumstances in which inequities are hard to track.

The "voices in the room" argument doesn't end there, either. Recent years have seen an uptick in large companies using software to break the law. From 2014 to 2017 Uber used a sophisticated application known as "Greyball" to operate illegally in hundreds of cities. In 2016 an online software and insurance company called

Zenefits was caught deploying a tool to help sales agents fraudulently obtain the licenses required to do their work (licenses designed to foster confidence among the public). Most egregious of all, between 2008 and 2015, Volkswagen ran concealed software to mask the fact that eleven million diesel engines advertised as "clean" were producing carcinogenic nitric oxide at four times the European Union limit. According to international research coordinated by MIT, Volkswagen's actions are likely to cost sixty US and 1,200 European citizens up to a decade of life. When it comes to killerware, this was the real deal, and a VW software engineer was the first conspirator sent to jail. Similar revelations appear by the week.

I can't prove any of what follows, but it seems self-evident to me that a room full of engineers of varied family and educational backgrounds, with a mix of ages, genders, races, classes and experiences of life—of casts of mind and personality—is more likely to contain one or more individuals moved to do what engineers in the above schemes did not do: raise their hands and say, "No! We can't do this, it's not right." To be sure, a team of young, privileged, single "compulsive programmer" men who've been schooled to consider breaking things sexy *could* contain such a person, but this is not the demographic one would choose to favor such an outcome. Could this be why some employers, especially those engaged in morally ambiguous activities, have been so eager to harness this narrow cohort? After all, it is, to them, *convenient*.

The curse of homogeneity is that problems and anomalies are less likely to be noticed than they would be in a more prismatic environment. Constraint of vision compromises decision-making in any realm, but in the domain of code, which is hard to do well, easy to abuse, and little understood by most of us, it represents a clear and present danger to society.

Catch 32

Computers are useless.
They can only give you answers.

—Pablo Picasso

I feel like a guppy swimming through Jell-O. This happens regularly. One week I'm making progress, enjoying the challenge of learning functions, conditional statements and loops, starting to feel my way into wilier concepts like classes and recursion, only to feel my skills melt away the moment I try to apply them, feeling ever more that I don't belong here; am in the wrong place with the wrong kind of brain or temperament. Worse, my projected big dream project, a functioning website, is sacrificed to practicality when I realize I don't have time to learn Django *and* assemble all the material I need. Tail between legs I engage a developer, a friend of a friend who is in his early twenties, plays standup bass in a jazz band for fun and does a bang up job in a few undramatic weeks. Moreover, I make the rookie mistake of failing to ask what languages will be used, which is how I now find myself needing to brush up not just my HTML and CSS to maintain the site, but learn some PHP as well. I tweet my Py-chums for advice on a PHP tutorial book and the response aligns neatly with what I would have expected had my plea been for a concise primer on seal clubbing. Even Nicholas Tollervey says "I generally avoid PHP," which as warnings go is equivalent to anyone else planting a horse's head in your bed.

There is a project of sorts on the horizon. In the months since PyCon, Nicholas and I have stayed in touch and begun to grow close. Even so, I was surprised and a little terrified when he issued an invitation to join him for PyWeek, a biannual code competition in which teams compete to make the best game in the space of a week. When I fretted about holding him back, he assured me otherwise. A friend of his had developed a pared-down Python game library called PyGame Zero, so this would be a chance to see how I, the consummate and possibly eternal beginner, responded to it in the field. Besides, there would be plenty of extraneous things for me to do, like source or create sound effects and music. In a frenzied week of fun and creation we cobbled together what amounted to a lo-fi version of the old classic *Frogger*, in which a player attempts to guide their frog across a road without getting squished. There were differences. Our frog became human, not for the conventional reason of having been kissed by a princess, but because Nicholas thought it might be interesting to have zombies prowl our sidewalks and zombies are known to seldom bother amphibians. Given the competition theme of "flow," we called our effort *Traffic Flowmageddon* and it was cheesy as a dungeon of brie. Faced with the staggering ingenuity of some of the competition, we in no way begrudged placing last in a field of thirty-one teams when the votes came in. But, hey, we'd finished; had made a playable game in the space of a week.

PyWeek had been madcap and electric. My favorite entry, built on a ravishing hand-drawn black-and-white landscape with snowflakes and meteorites drifting from the sky, placed third and comfortably ascended to the level of art, as did many of the top entries. Better still, all the code on view was ready to be shared, copied, tinkered with or rewritten by anyone—unbelievable! I contributed little code, concentrating on sound, but felt privileged to have breathed the same virtual air as these creative game maestros, and I happily signed up for a rematch six months hence.

What did I learn from my first PyWeek? A little more about constructing algorithms and the uses to which they can be put. To my surprise, the same awkward sequential, binary logic I've found so uncomfortable—that something in me has *wanted* to resist, I suspect—showed the first potentiality to be alluring and even fun. One garden variety algorithmic trick I found pleasing in *Traffic Flowmageddon* was the technique used to give zombies their trademark jerky movement, the like of which must be as ubiquitous to the microcosmos as bacteria to nature by now. In our version Nicholas imported graphics of a zombie in five different positions, each like a movie still of somebody walking.

These images sat outside our code in an images directory, ready to be grabbed and displayed as needed. Think of them as "frame 1" through "frame 5." Elsewhere in the code a "make_zombies" function randomly generated ghouls and placed them in a list labeled "zombies." The zombies list would contain those monsters currently active in the game. If three zombies were in play, the list might look like this:

```
zombies = [zombie1, zombie2, zombie3]
```

Now Nicholas set a *for loop* to iterate over our zombies list, advancing the image being displayed for each active zombie by one frame with each pass, as in a film, producing the creepy lurch familiar to horror movies and venture capitalist conventions. The business end of the function written to automate this task appeared thus (with explanation to follow):

```
1 def animate_zombies():
2     for zombie in zombies:
3         zombie.frame += 1
4         if zombie.frame == 5:
5             zombie.frame = 1
```

What's happening here? As before, *def* means "define a function," which in line one we are naming "animate_zombies." As we know, functions in Python are identifiable by parentheses at the end of their name. In a function *definition* such as this, the parentheses stand ready to be filled with *parameters* specifying the sort (or sorts) of data input the program expects to be given to work with. For example, a function for adding names and addresses to a dictionary will expect a name and address to be supplied whenever it is *called* into action. In this way, the parentheses become the portal through which we provide *input*—the stuff we hope to see transformed. Computerists speak of *passing* these input data to the function as *arguments*. The word "argument" made me laugh at first, given what I was learning about coders, "religious wars," etc., until I learned of its root in a mathematical term dating back to at least the fourteenth century and meaning "a thing from which another thing may be deduced." In the case of "animate_zombies," which merely iterates over a preexisting list, no fresh input is required, so the parentheses remain empty.

In line two we establish the *for loop* that does our work. The Python interpreter always takes the word immediately following "for" to represent individual items in the *collection* we want it to iterate over (in this case the zombies list). Our choice of the word "zombie" for this purpose means nothing to Python, the interpreter responds only to position within a rigidly defined syntax: "Zombie" is usefully descriptive for a human reader, but we could have used any signifier—"zombies_rule," "hairdo," it wouldn't matter. The interpreter reads line two, with its colon at the end, as "for each zombie in the list 'zombies,' do the following." This is the essence of iteration.

Line three tells the interpreter *what* to do with each zombie it pulls from the list to act upon. In Python, everything indented after the colon is included in the loop. When I first saw the word "frame" linked to the "zombie" object by a period in this line, I assumed it to be a *method*, the species of function that attaches directly to an object and makes it do something. Nicholas pointed out that there were no parentheses after "frame," indicating that this was not a function but an *attribute*—a property pertaining to the object, just as a person might have the attribute of green eyes or tiny feet. In this instance, each zombie object is given a corresponding "frame" that points to one of the five images in our images directory. The "+=" *addition assignment operator* means "add 1 to" or "advance by 1," so "zombie.frame += 1" means "move from the current zombie image in the image directory to the next one." *Voilà* lurchy gait.

But in lines four and five Nicholas lays a condition on the loop. Because only five images are available in the image directory, when a zombie in play reaches "frame 5," we need it to go back to the beginning. We know that in computing the mathematical equals sign is used as the *assignment operator*, simply to assign values to variables ($x=15$, for instance). "Equal to" is thus denoted by a double equals sign (==), which performs a Boolean test for equivalence, returning *True* if the objects on both sides of the sign *are* equivalent and *False* if not. In line 4 we use this symbol to set the condition. If our zombie has reached the frame corresponding to image 5, then in line 5, we ask the program to reassign it to frame/image 1 and continue looping for as long as this creature is active in the game—that is, remains in the zombies list. In the program, the "animate_zombies" function occupies lines 306–313, and I think it's cute.

🐞 🐞 🐞

Yet the revelation of PyWeek for me was GitHub. Anyone who has spent time learning to code recognizes a scenario in which a program is progressing well when a small change breaks something; a fix for the change breaks something else; a fix for the fix breaks everything and the horse it rode in on, then kicks it all the way to Albuquerque. To spare this perdition, programmers use a technique called *version control* (or *source control*), which allows work to be logged at intervals and changes to be rolled back. Ubiquitous to this process is a site called GitHub, whose prime value is in managing source control for teams.

GitHub is one of the most important points in the microcosmos. And one of the most befuddling. It allows contributors to a project to suggest changes or additions, to be vetted and accepted, adapted or rejected—later reverted if necessary—with agreed refinements sent to the local machines of everyone involved, keeping participants on the same page at all times. It does this by providing *project repositories* (central locations for storing program or project files) from which code may be *cloned* to would-be contributors' *personal repositories* and thence their personal computers, to be worked with. I could see that GitHub's artistry was to choreograph a great skirt-swirling waltz between these locations, but the question of how was beyond me. I hit the tutorial trail, only to encounter a deeper and more general hitch, which, with apologies to Joseph Heller and the hexadecimal memory addressing system, I came to think of as a "catch-32": *if you can understand the tutorial, you don't need the tutorial.* GitHub's logic could make a Minotaur blush.

Not for the first time, the day was saved by a women's tutorial site, Women 2.0. Here I received the priceless information that the coordinator of this source control dance was not GitHub itself, but an unassuming tool that sits among the apps on a laptop and makes the whole thing work. The tool is *Git*, which is where GitHub gets its name: it is a hub for distributed gits.

Git? In British English, where "git" is synonymous with "asshole," a "hub of gits" could pass for an insult dear to the innovative swearers of Scotland, perhaps after that versatile Irish classic, "a shower of bastards." With pleasing symmetry, the Git application turned out to have been conceived by none other than Linus Torvalds, the ego-rich coding legend behind the Linux operating system and its harsh development culture (until 2018 Linux developers had not a Code of Conduct, but a Code of *Conflict*). And this is where source control gets interesting. Like so many of the best software tools, Git began as an incidental adjunct to a larger project, when in 2005 Torvalds wrote a small program to allow Linux developers to coordinate their work. So useful was his Git app that it mustered its own vast hinterland, of which GitHub is part. What this meant was that before I could use GitHub, I needed to install Git. This done, the dance began to make sense and I gained at least toddlerish control.

An intoxicating universe yawned open, because GitHub is the locus of code's profoundest operating principle, the open-source software (OSS) notion that collective, distributed intelligence always transcends the sum of individual minds within a group—and that time gifted the community is well spent because a rising tide lifts all boats. At the end of 2022, site administrators claimed more than eighty million users and two hundred million project repositories ("repos" in gitspeak), roughly half of them public. Among the latter is the source code for projects as varied as Bitcoin; freeCodeCamp; most programming languages; mathematician John Conway's thrilling evolution simulator, *Game of Life*; the TempleOS operating system, which developer Terry A. Davis claimed to have been visited upon him by God. The German government trusts GitHub with the text of its federal laws. *Traffic Flowmageddon* would be safe there.

I found that the site also hosts illicit code, including game console emulators, exploits purloined from the National Security Agency by the hacker group Shadow Brokers and tools for circumventing Chinese state censorship of the internet. How many of these repos would survive Microsoft's purchase of the site in 2018 was unclear, and yet GitHub remained the rare example of a high-profile American website that almost never got hacked, because the bad guys rely on it like everyone else. Often claimed as being to the digital realm what the Library of Alexandria was to antiquity, GitHub's loss would be similarly catastrophic. Elated to have gained entry, I went on an orgy of project surfing, "cloning" whole repos to my hard drive and into my editor, a process that felt like a journey through physical space, a digital road trip setting out from my own website and ending—deliriously—with the assembly language source code for the Apollo 11 lunar module that got Neil Armstrong and Buzz Aldrin to the moon and back in July, 1969.[1]

---

1 Anyone can do this: You can do it now. Go to GitHub and search "chrislgarry/apollo-11." Click on the files to see the code. Or better still, clone the files and open them in Mu or any open-source code editor.

Especially intriguing were the politics. Larry Ellison of software giant Oracle is reported to have deemed the open-source movement "un-American," while Microsoft's Steve Ballmer branded Linus Torvalds's talismanic open-source operating system Linux communist. Yet, while one of the key articulators of the open-source ideal is the computing book magnate Tim O'Reilly, a thoughtful denizen of the democratic left, another, Eric S. Raymond, author of the excellent open-source set text *The Cathedral and the Bazaar*, is an activist of the ornery, gun-rightsy, #MeToo-bashing libertarian right. This intellectual fuzz heightens the movement's exoticism, especially once we introduce the fascinating question of why individual coders choose to contribute as they do rather than sit back and reap the rewards as fellow travelers. On first contact with the open-source ideal, I'm staggered. Have coders hit on a way of being that shows all of us a way forward?

Merriam-Webster defines "movement" as "a series of organized activities working toward an objective." In real life, such "objectives" nearly always exist in opposition to something else. It takes me a while to grasp what the open-source experiment, this seeming digital Jerusalem, is reacting to and why it is so important.

Every fifteen years or so a particular story appears in the British press. You stop being surprised once you've seen it come around a few times, but like an astronomer greeting the return of a favorite comet, you might smile when you see it anyway.

Back in 1973, the year of Watergate and the gas crisis, Ziggy Stardust and *The Exorcist*, the British Railways Board was granted a patent for a flying saucer. Last time around this story elicited not just smiles among the public, but an unanticipated pang of nostalgia for the good old days of the state rail monopoly and an age when managerial ambition extended beyond shareholder value and goosing executive pay on the sly. The old British Rail might have struggled to get you to Cricklewood on time, but that was okay because (a) it was still better than the extractive private monopolies running the show now, and (b) it had plans to take you to Mars.

I'm thinking about this because I'm thinking about the motion of technology. Viewed in isolation and from fifty years' distance, the Railway Board's plan looks ridiculous. It had asked an inventor named Charles Osmond Frederick to design a "lifting platform," only for him to return with plans for a clean, super-fast, space-ready saucer. According to the original patent document, the vehicle was to be powered by laser-controlled thermonuclear fusion, a technique Frederick might reasonably have presumed imminent but which, like so many expectations of that era, remains a twinkle in engineering's eye at this writing. So eventually the patent lapsed and the saucer was consigned to its archival graveyard, waiting to tickle future generations with the idea that an organization unable to serve drinkable coffee on a train might contemplate reaching for the stars. Look a little closer, though, and the British Rail saucer emerges as something far more intriguing and substantial, telling a broader and more nuanced story about our relationship to our machines; to innovation; to software.

A few years back, the novelist Ian McEwan noted the way science "prefers to forget much of its past," being "constitutionally bound to a form of selective amnesia." Redundant theories and failed designs have no further utility and are quickly buried, he said, paying a definitive price for being wrong. And one of the places these ghosts of futures past are buried is in the Business and Intellectual Property Centre at the British Library in London, where an archive of more than sixty million patent specifications from forty countries are kept, dating back to 1855. All searchable remotely.

Enter the archive and a seldom-seen wilderness of fantasy and folly opens before you. Within this bureaucratic dreamworld, the first shock comes almost immediately: galaxies of patents for flying saucers were filed around the time Charles Osmond Frederick filed his, in the optimistic wake of the first moonlandings and maiden flight of the supersonic airliner Concorde. Frederick was part of a generation of mechanical engineers for whom anything seemed possible. How could they have known that our technology, mirroring—or was it driving?—our collective psyches, was about to turn inward and *shrink*?

Dig deeper into the archive and these legions of lost patents coalesce into a secret social history. Even the absurdities tell coherent tales if you look closely. An "airplane hijacking injector" (US3841328) involved stationing hypodermic needles under every aircraft seat, ready to "sedate or kill the passenger" at the flick of a pilot's switch. In the ecstasy of creation, it never occurred to inventor Jack Jensen, of Fort Worth, Texas, that the rest of us might choose to paddle the Atlantic before braving seats equipped with death needles. All the same, hijacking was a big issue when the patent was filed in 1972 and proposed solutions are scattered throughout the records, clustering in the early 1970s and after 9/11, just as private nuclear fallout shelters and esoteric aids to domestic life litter the 1950s (try "sanitary appliance for birds," GB2882858) and increasingly desperate barriers to HIV define the late eighties. It should come as no shock—and yet does—that the one constant is sex, from Victorian methods for preventing "nocturnal emissions" and "self-abuse" and encouraging chastity in the manner of Mrs. Ellen E. Perkins's "sexual armor" (US875845), to the forests of novel twentieth-century condoms, ranging from the Femidom to a "force-sensitive, sound-playing" number capable of elevating the moment with *God Bless America* or something by Miley Cyrus.

When historians look back on the present moment, they will see a tsunami of patents related to AI and machine learning. A trawl through the 110 million patents in the European Patent Office's Espacenet Global Patent Index reveals a large majority of these entries to be from just three countries: China, Korea and the United States. Back in Berlin, I wondered why there is no European software industry to rival Silicon Valley. Now I find out why, when the patent and inventions specialist Stephen van Dulken tells me something I feel I should have known but didn't. Historically, the United States is one of very few countries in the world to grant patents for software. Hence Google. Hence Microsoft, Facebook and Amazon. Hence nearly all major software companies in the West being American. In common with most of the world, Europe broadly adheres to Dijkstra's view of algorithms as part of the fabric of the universe, waiting to be *discovered* rather than invented, more akin to mathematical proofs than Mrs. Perkins's sin-squelching armor.

As Van Dulken walks me through some landmark filings from the archive, the gray areas underlying this divergence start to take shape. There is no clear consensus on when the first software patent was granted, but the maiden recognition of an algorithm was in 1965. Martin Goetz's "sorting system" (US3380029) did just what it said: organized large numbers of files to be stored on tape. The same year, Goetz crossed a notional line in the sand when he filed and brought to market the first piece of commercial software, a program for printing flow charts (US3533086). Important to note is that in keeping with practice ever since, no specific lines of code appeared in Goetz's specifications: the pages contain sketches that describe the general logic of his algorithm. For the US patent office, originality of *concept* has always been the point of a patent. In my earlier conversation with James Dyson, the inventor chuckled as he explained that the distracted mad scientist inventor is an officially recognized construct.

"You don't get a patent unless you happened across it unintentionally," he told me. "It's an invention: you come across it, you don't deduce it from discernible facts. If it could have been done by one 'skilled in the arts,' you don't get a patent. That's why there are so many apocryphal stories about how inventors happened across their inventions while looking for something else—because you have to make them up in order to get a patent!"

The idea of epiphany, of the shamanic techno-seer touched by some unseen force of invention, is written into the process. Dyson pointed out that established patents in the same field as a new application are referred to as "prior *art*." Code patents may be harder to sanction simply because nobody ever discovered a function while searching for a headache remedy or had an algorithm fall on their head from a tree. Code can be copyrighted in the same way as books or pieces of music, but copyright provides only limited protection.

And this is where the ownership issue gets complicated. More limited protection is what many experts have wished to see for software, because the effect of patents—in contrast to copyright—is to block off whole avenues of innovation by rendering them proprietary for (in most cases) twenty years. The open-source movement rose explicitly to prevent such outcomes. For similar reasons, software patents are granted in the United Kingdom only if they contain algorithms necessary to the working of a physical invention, and in most of Europe under even more circumscribed conditions.

So. To protect or not to protect? As with most code issues, the answer depends on how you frame the question. Scroll forward from 1965 to the first gasp-inducing patents of the internet era and the implications of our choice become clearer. In 1998, at Stanford University, one Lawrence (Larry) Page received a patent for the "Page rank search engine page ranking system" (US6285999) that would instantiate Google. Page and his fellow doctoral student business partner Sergey Brin were the sons of academics: Page's father was a computer science professor and his mother also taught programming, while Brin's mother and father, after negotiating a tense exit from Soviet Russia in 1979, worked for NASA and taught math at the University of Maryland, respectively.

Valley libertarians seldom advertise the fact that Brin was at Stanford on a National Science Foundation fellowship, so government lurked modestly in the background at the inception of Google (as per so much innovation attributed to the private sector). The drawings in Page's patent describe algorithms adapted from those used to rank papers as part of the academic citation process he and Brin watched their fathers use. Programmers smile when they tell you they've heard the Java and Python code written by the duo was clunky, and that early Googlers referred to a part of the program called BigFiles as *BugFiles*—though no one disputes that it did the job, at least until it didn't and was rewritten by the prodigious "pair programming" team of Jeff Dean and Sanjay Ghemawat from 2000 onward.

A year earlier, in 1997, Amazon had outraged webbies by applying for and being granted patent protection for a "1-Click" ordering system that removed the final layer of friction from their online purchase process, then successfully suing Barnes & Noble for adopting a similar system. As with Google, Amazon's patent document supplied algorithmic architecture rather than code and laid the foundation for a globe-straddling US company that couldn't have originated anywhere else. In the years between the granting of these foundational patents and their recent expirations, it would be true to say no one came up with better one-click ordering or search systems than Amazon or Google. *But.* This is not because the technical ability to improve on the originals didn't exist, it is because Amazon and Google owned the *general concept* of how the necessary algorithms operated . . . leaving us with a pair of monopolistic tech behemoths that stifle innovation and have started to scare us (trust me when I say that if you're an author Jeff Bezos doesn't just *look* like Dr. Evil from *Austin Powers*). The situation is not unlike a construction company having been granted a patent for "Physical structure with roof, walls and windows," which would certainly have made a few people very, very rich while consigning most of us to huts and caves.

If only the issue ended there. When Van Dulken tells me about a tableturning 2014 Supreme Court ruling in the case of *Alice Corp v CLS Bank International*, which led to software applications of a more "abstract" nature being rejected and protection becoming far harder to establish—with some previously granted patents revoked—I went away assuming this to be a good thing. A perusal of court documents later revealed that among parties arguing the case before a bench of flustered Supreme Court justices was the Electronic Frontier Foundation (EFF), ardent campaigner for an open and free digital realm. For the EFF, tightening the criteria for protection of algorithms promised to stop "egregious software patents" filed by so-called "patent trolls."

"To take a typical example," the net watchdog noted in response to the Alice ruling, "a patent troll called DietGoal, armed with a stupid patent on 'computerized' meal planning, sued a bunch of websites simply for providing online recipes or menus. . . . Today's *Alice* decision provides a tool for courts to throw these cases out."

Who could object? Until an email exchange with Rana Foroohar, the New York-based associate editor and global business columnist at the *Financial Times*, alerts me to her book *Don't Be Evil*, in which she points out that the most strident advocates of the 2014 tightening of patent law were Big Tech lobbyists. With the twenty-year patents on which they built their monopolies set to expire, it suited them to make protections harder to establish and defend, especially for startups without the financial and legal muscle to resist challenge. This change was bound to have a chilling effect on the tech sector, and it did, leading to a "dearth of startups, declining job creation, falling demand," according to Foroohar—a verdict later confirmed in an excoriating report by the US House of Representatives. Why pour energy into innovation if a tech giant is likely to co-opt your idea the moment it shows promise? And why would a venture capitalist invest in genuine innovation under such circumstances? One answer is "to be bought by one of the giants as a defensive measure," which severely limits the scope of interest, is both anticompetitive and dull. Yet the House report notes that even this minimal consolation is disappearing, with Amazon rebranding and selling open-source software as its own, forcing software creators to adopt more restrictive licensing arrangements and benefiting no one at all. No wonder long-standing free market fundamentalists like the Nobel Prize–winning economist Paul Romer now decry Big Tech monopolies as inimical to innovation and want to see them regulated.

This is the backdrop to any discussion of code and software in the twenty-first century, and it does play out in real life. One thing I notice from living in the penumbra of Big Code in the Bay

Area is a broad disparity between the gilded lives of tech friends and acquaintances lucky enough to work for the FAANGs (Facebook, Amazon, Apple, Netflix and Google), plus a handful of others, who tend to be secure and extremely well paid, and the frequently hardscrabble precarities of those who don't. At the heart of this issue remains the question of what software actually is—and the truth is we haven't decided, and aren't even sure where it ends and we begin. Open-source software development cast itself like a spell into this gap.

Open-Source Software (OSS) had a more radical predecessor: the *free software movement*, initialized circa 1983 by another divisive coding figure named Richard Stallman.

As a young computer scientist at MIT in the late 1960s, Stallman was mentored by original members of the storied Artificial Intelligence Lab, a group of prodigious misfits immortalized in Stephen Levy's book *Hackers: Heroes of the Computer Revolution*. They are also thought to be the inspiration for Joseph Weizenbaum's warning about the dangers of "compulsive programmers." Rapt at the vision of a community in which knowledge and information could flow free as air, Stallman took the "hacker ethic" to heart, seeing within it a "constructive anarchism" that could replace the "dog-eat-dog jungle" of mainstream American society with "a concern for constructive cooperation." While the Vietnam War was in full swing and America's future in the balance, he was not the only one to see freedom in a form of radical sharing, but he took the ethic to extremes, waging war against passwords at MIT in a way that could equally appear socialistic or imply limited understanding of and empathy for people as they actually are (a traditional conservative might say these amount to the same thing).

Later he would lead quixotic demonstrations against commercial software companies such as Lotus, never wavering in his view that software should never be owned. None of his substantial contributions to computing belong to him or anyone else. In 1991 he received a coveted MacArthur Foundation "genius" grant. So significant have his contributions to software been that, as per Dijkstra, he is often referred to by his initials alone: RMS.

Like Linus Torvalds, RMS also exemplifies a contradiction we see often in code lore: someone whose stated MO is cooperation, sharing, communication, yet who strains to connect with others on an individual basis. By his own account, RMS has spent most of his life in pain, feeling lonely and isolated. The preface to a book he wrote in 2002 could speak for everyone I meet who knows him when it says, "I don't know Richard Stallman well. I know him well enough to know he is a hard man to like." In the afterword to a 2010 edition of *Hackers*, Stephen Levy found the free software maven feeling like the last member of a dying tribe: the one true hacker left on Earth. "I'm the last survivor of a dead culture and I don't really belong in the world anymore," RMS said. "And in some ways I feel I ought to be dead."

The free software movement and its more pragmatic cousin open source, with whom it has had an at times strained relationship, share a common point of genesis. As told by Eric Raymond and Tim O'Reilly—in whose Oakland garden I spend an entertaining morning that includes a discussion of the open-source ideal—it all goes back to Unix.

"Unix" is one of those words most of us have seen in passing but thought too boring to pursue, like "exec file" or "http." News that it pertains to a computer operating system may not help with this. And yet when one of the industry's star figures, Ken Thompson, wrote Unix in the year of Apollo 11, ARPANET, *Abbey Road* and Woodstock—*1969*—his innovation was profound enough to inspire another seminal computerist, Dennis Ritchie, to conceive the Sanskrit of modern programming languages: *C*. Why did this matter? Because until then, operating systems tended to be written in terse assembly code in order to spare expensive, limited memory. Because each model of computer chip processed machine code in its own way, the assembly language would be specific to that chip, meaning the operating system would have to be too—none were transferable between different types of machines. This in turn meant each program had to be written or rewritten for the machine on which it was to be used—and when obsolescence struck (as inevitable for computers as death and taxes for people)—the entire software ecosystem had to be recalibrated. This was just the way things were: finnicky and irritating.

Thompson and Ritchie saw something few others had. Hardware and the compilers that translated between higher-level languages and machine code had improved to such a degree that an operating system could now be written in C. This would use more memory, but open the way for the system to run on any machine capable of running C itself. When a new model of computer chip was introduced, the manufacturer would need to write a bespoke new C compiler for it—a skilled but relatively straightforward task—after which not just the operating system, but any software composed in C could run. Thereafter, a universal operating system could serve as a standard environment for programs of any size and complexity on any machine. Software would be portable. Programmers would need to learn a new program only once and could carry favorite tools around with them. Thompson and Ritchie's operating system would be called Unix, an unsexy name for a very sexy idea—if you were a programmer. There have since been dozens of Unix based ("Unix-like") operating systems, the best known of which are Apple's macOSX and Android.

Both Unix and C caught on fast. But there was a catch. Having begun at Bell Labs, Unix was proprietary, meaning *owned*, and by 1984 it belonged to AT&T. Antitrust laws forced AT&T to license their operating system to whoever wanted it ("What about Amazon and Google?" we may ask), but in practice this limited its reach to institutions. Individual computerers were excluded.

Enter RMS. The free software movement and its formal expression the Free Software Foundation arose specifically to develop a people's version of Unix, but the project was never completed. In 1991, Linus Torvalds grabbed the baton and finished the job, not because he was 10Xier or more capable, but because he had a tool Richard Stallman did not. The World Wide Web, launched by Sir Tim Berners-Lee that year (to less fanfare than the contemporaneous introduction of small plastic balls into cans of "draught" Guinness) opened the possibility of recruiting the whole world as a development pool. An enthusiastic ad hoc community formed and—in a radical break with everything anyone thought they knew about project management or the creative process—got the job done in short order. In Eric S. Raymond's words: "The Linux community seemed to resemble a great babbling bazaar of differing agendas and approaches . . . out of which a coherent and stable system could seemingly emerge only by a succession of miracles. The fact that the bazaar style seemed to work well, came as a distinct shock." Raymond continues, "Joy is an asset. It may well turn out that one of the most important effects of open source's success will be to teach us that play is the most economically efficient mode of creative work."

A notion technologists very much took to heart. On the face of it, Linus Torvalds had discovered—an anthropologist might say rediscovered—something important about Homo sapiens. Given a clear and worthy goal, people would go to work for a community; could be self-motivating and spontaneously self-organizing within a minimally designed social framework. In a decade, the 1990s, when corporate managers and bankers were selling the idea that only vast bonuses could generate the sort of commitment they needed to do their jobs, and "pay" was being reframed as "compensation," Linux, like the open-source model it pioneered and the web itself, looked like a utopian revelation: proof that a better world was possible—perhaps inevitable. What was more, no one owned the source code. This innovation belonged to everyone.

🐞 🐞 🐞

Perhaps the purest expression of open-source spirit in action is the coder's help site Stack Overflow, which is why I am sitting in the New York office of Jay Hanlon, the company's present executive vice president of culture and experience, trying to figure out how and why the paradigm works. I am not the first to wonder. Hanlon smiles as he rehearses a Thanksgiving Day conversation his colleagues all know by heart.

"Typically, you have an aunt or an uncle," he begins, "and they ask, 'So what does your company do?' You explain it and they say, 'Okay, but who answers all the questions . . . do *you* answer the questions?' No, people do. Then they say, 'Well, how do you pay all the people?' And you say, 'We don't pay them, they just do it to help each other.' And there's always a pause, as they try to figure out if they're missing something and realize they're not. Then they just say: '*Why?!*'"

In auntie's defense, this is not a stupid question. If Facebook profits from harnessing the noxious side of human psychology, as a kind of wind turbine for psychosis, Stack Overflow appears predicated on all that is noble. Not a business model for the faint of heart, you might think. More broadly, these two giant sites represent the yin and yang of what code can do, the first looping avarice into a chaotic ball of rage, the second channeling generosity in a way that can't help but gratify by contrast. Now more than ever, the site asks important questions about people in society, about how such a community sustains and why individuals opt to sustain it with apparent selflessness. Can this be what it seems? And if so, could it be replicated elsewhere?

Hanlon came to New York to be a playwright, but in the New Yorkiest way imaginable took some temp work to pay the rent and woke one day to find himself selling options at Merrill Lynch. Fourteen years later he was coaxed to the fledgling Stack Overflow, where I find him overseeing community affairs with amused good humor and heart. A glib answer to the question of how contributors stay motivated—one he hears all the time, he tells me—is "gamification," in the form of a carefully constructed chain of feedback aimed to bolster constructive engagement. A user of any skill level posts some code they can't make work. Peers return answers, which others with history on the site can up- or down-vote based on perceived helpfulness. More can comment on both the original question and answers, offering refinements and corrections where necessary. At every stage, positive contributions lead to positive feedback—clever, to be sure, but as Hanlon points out, gamification can only reinforce behavior people want to engage in (such as playing games). Stack Overflow answers take thought and often considerable time to formulate: few individuals would craft them for brownie points or a fleeting buzz alone.

Hanlon's view of what drives his community is at once more complicated and pleasingly homespun. By default, most people like to help others and will do so absent a reason not to, he suggests—and science tends to agree. Our evolutionary toolkit includes a neurotransmitter and hormone called oxytocin, which appears to foster and reward connection to other people (although this does not necessarily make oxytocin *moral*, as has been claimed). A feeling of achievement in solving difficult problems can also be a factor, Hanlon thinks, but most important of all might be the community dynamic itself, a sense of being part of something bigger than oneself.

"So yes, the Stack Overflow system lets you help people, lets you solve challenging problems," Hanlon summarizes. "But the other thing it does is promise that if you can do something useful here, you have created an artifact that is part of this big system. And whenever anyone else has this problem, they will come and find your answer."

In time I will know and understand the power of this sensation.

Only when you ask why other groups don't emulate Stack Overflow do you start to realize how unique coders are as a group. Most professionals maintain status by guarding knowledge (lawyers, accountants), or need explicit credit for their work to attract future funding (academics, writers, musicians). Medicine might be a candidate, except that facilities, practitioners and patients are dispersed and to a large extent specific. But where a brain surgeon is on their own in an operating room, no coder is ever alone in the web age. Modern programmers are collaborative by necessity and accustomed to using and building on each others' work. A culture of sharing obviously benefits the majority.

Two dangers lie in wait for a large forum like this, Hanlon explains. The "Yahoo Answers problem" arises where the acceptable subject matter grows so broad that no one knows much about anything, leaving users to scroll through pages of unhelpful answers before finding anything useful. But the converse can also happen.

"For forums that don't fall into the trap of getting too broad, the danger is that they go the other way over time. New people are seen as the irritant, the threat—they're the ones that are going to ruin everything. And what you wind up with is, instead of being the 'We Collect Smurfs!' forum, they're the grumpy old guys who don't like the new people who don't even know a damn thing about Smurfs yet but have the nerve to ask a question here . . . so you wind up shrinking to the point where you just collapse in on yourself."

For this reason a set of firm, community-enforced rules are employed to keep exchanges focused and relevant. Hanlon points out that no other specialist community website has reached Stack Overflow's size and maintained what he calls a "stable orbit." Management hasn't always given clear guidance to newbies on how to engage with the rules (post your code, explain the problem, describe any fixes you've already tried, don't raise issues you could solve with a tutorial), leading to some withering slapdowns. But such insensitivities are now being acknowledged and addressed. Hanlon refers me to a recent blog he wrote, laying out the issues ("too many people experience Stack Overflow as a hostile and elitist place . . .") and proposing remedies, which look well thought out and proportionate. Like the rest of civil society, the programming firmament is transitioning from a narrow view of its history and purpose to a more nuanced and modern one informed by movements like #MeToo and Black Lives Matter. For some, this is uncomfortable—as change typically is.

"I think there was a time when the people who were drawn to programming, at least in part, were people who were . . . how can I say this? One of the nice things about programming a computer is you tell it what to do and you don't have to worry about *how* you tell it—if you tell it the most efficient way, it works. And talking to humans, the most efficient way is often not the way to get the best results. We've always had a policy that says *you can't be a jerk*. Ad hominem attacks are never okay, name-calling was never okay. It's sad that this should be a source of pride on the internet, but in twenty-five million user-created Stack Overflow pages you won't find a single racial epithet or homophobic slur. And it's not because we're so amazing, it's that when horrible stuff happens, a user will always say this isn't cool and tell a moderator, who deletes it almost instantly. But what we're trying to get better at is the softer stuff, training people to understand that, when you're conveying something along the lines of 'You just need to use this basic approach,' to maybe take out the word 'basic.' So much of this is about little things where you can help someone without making them feel dumb."

Empathy, in other words. To my surprise, Hanlon doesn't think the general deterioration in civil discourse has directly affected Stack Overflow, mostly because the focus remains squarely on programming. The culture war in code is more epistemological, it seems to me, revolving around *feelings* and the question of whether emotion has value and empathy validity in the realm of code—or whether some imagined "pure reason" should be the engine and arbiter of all interactions and outputs. Either way, Stack Overflow appears to show that a balance can be found, and I reenter the Midtown Manhattan fray viewing the open-source miracle with something close to awe. If only the rest of the world could be this way, I marvel. It seems almost too good to be true.

CHAPTER

10

A Kind of Gentleness

She said gently that they believe when a lot of things start going wrong all at once, it is to protect something big and lovely that is trying to get itself born—and that this something needs for you to be distracted so that it can be born as perfectly as possible.

—Anne Lamott, *Traveling Mercies: Some Thoughts on Faith*

In retrospect there were hints, not least a much-discussed PyCon 2018 talk by a Python core developer named Brett Cannon, who described having to withdraw and seek emotional support in the face of abuse from open-source contributors. I'd missed Cannon's talk, assuming his complaint to be insider stuff, the push-and-pull of a passionate community. But this was a Sunday morning keynote: someone thought it important. Now I see why, as the tinderbox he described ignites close to home and with devastating effect.

One of the first decisions a would-be programmer faces is selection of a code editor, a choice that should be easy but is not. Pros frequently use one of a rarefied open-source pair respectively called

Vim and Emacs—the latter owing its genesis to Richard Stallman—but these are feature rich and take patience to set up, a poor use of time and finite pain reserves for a beginner, like deciding to master violin as an aid to learning cello. Beyond Vim and Emacs, no two coders have recommended the same editor in my presence from among a perplexing panoply that grows by the day. So, when Nicholas announced his intention to launch a friendly open-source editor aimed at beginners and those who support them, no one needed less convincing of his idea's worth than me.

Nicholas named his editor Mu and it fulfilled its brief in style. As with all software, there were bugs to be captured and returned to their natural habitat among the odd socks and lost pens, but none that seemed to matter. I continued searching for the pain-in-the-assy grown-up editor I knew I would need but found myself gravitating ever more to Mu for practice, while Nicholas gathered a community of open-source developers and threw himself into the project, excited and having fun. Soon after first release, someone tweeted a photo of a classroom full of kids using Mu—in China. Nicholas was overjoyed and I felt elated for him: he was on to something. Here was open-source development at its best, an example of how societies *can* be run.

The change came abruptly, or so it seemed to me, in a Twitter thread making clear my friend had been viscerally upset by something. The content of a new blog post he wrote took me aback.

> *Over the past three years I have grown despondent about the problems I have encountered in the UK's Python community. This year the feeling became unbearable, to the extent that I sought professional help to deal with mental health problems solely arising from my contact with the UK Python community. . . . Last weekend a straw broke*

the camel's back and, after a considerable amount of thought and reflection, I finally decided to publicly reveal how I felt via Twitter. I've been sitting on a huge amount of pent-up frustration and sadness, and there needed to be a controlled release.

The problem was "a lack of just one thing," he said. Compassion. As if to confirm as much, Nicholas's plea for humaneness within the UK Python community had been dismissed, attacked and worst of all *denied* by others. His consequent and understandable distress was grievous to see. How could anyone with an ounce of empathy fail to recognize that something had gone very wrong here—something a mature community should feel impelled to address rather than assail? A later tweet read: "*sigh* And now I've just heard news of two dear friends in the international Python community who are on the receiving end of Internet bullying and trolling. What is wrong with people . . . ?"

By coincidence I was in London, so shot Nicholas a message of support and we arranged to meet in the cafe at Foyles bookshop, where the depth to which he had been shaken by his experience with Mu and the wider Python community was clear. And mystifying. Even at this early stage of development the editor worked well. It was free to download and use, so Nicholas stood to gain nothing for himself save the educator's satisfaction in sending something of value into the world: a tool to declutter the learning process for beginners, removing an unnecessary barrier. *Hallelujah*, surely . . .

Nicholas talked and I listened, straining to make sense of what I heard. Snarky, carpy, mean comments. Ad hominem attacks: name-calling. Messages along the lines of "ten reasons your editor sucks," which would be cruel and unnecessary even if Mu did suck. But it didn't. Affronted gripes that the editor wouldn't work with an expert user's highly customized local environment, rising to full-bore anger when Nicholas apologized and explained that, at this stage, there wasn't time and energy to tackle a compatibility issue likely affecting one person in the world—but that, this being open-source software, his antagonist could always fork the code from GitHub and adapt it for himself. Cue rage. Bellyachers refusing to accept a decision and let it go, indulging in a practice called "sealioning," arguing in infuriating circles with no regard for the time and energy they were demanding of someone who, when all was said and done, was just trying to do something good. The behavior Nicholas described was appalling, inexcusable. No healthy federation could tolerate it. I admired his courage in speaking out.

Now a curious thing happens. Torn by the unmerited pain of a friend, I start to see signs of OSS's complicated shadow side everywhere: in a developer named Sage Sharp's anguished departure from the Linux kernel development community with a blog post very like Nicholas's; in a conversation with the brilliant Russell Keith-Magee, whose DjangoCon talk I'd so enjoyed, where he reveals that he, too, had been forced to pull back from Django core development and seek mental help due to hostility from would-be contributors (and this in a community whose official culture could not be more open and humane) . . . in word that Guido himself has suffered over the clamorous, contested adoption of the so-called "Walrus Operator" *to the language HE DESIGNED.* Further news that Linus Torvalds is on leave from the Linux community he founded after an internal rebellion against the solipsistic, pretentiously "meritocratic" culture he built, and that it took his daughter to persuade him to seek help in trying to "learn empathy." Noting both the pathos inherent in the concept of "trying to learn empathy" and insider doubt that reform is even possible at this stage, especially against the strident objections of some long-standing community members.

I go back and watch a recording of Brett Cannon's PyCon talk about a "crisis" in OSS and find the discussion so uncomfortable that I call him, hoping to understand this threat to what I'd grown to regard as coding's luminously beautiful exemplar.

To contribute to an open-source artifact such as Python, developers typically go to the project's GitHub repository, peruse a list of known *issues* and offer to work on one. Issues can range from spelling mistakes to bugs to aspirational features. Alternatively, the hopeful contributor can open an issue themselves and offer a solution. Once the work is done, they open a *pull request* (or *PR*) to have the suggested change reviewed by a core developer and either rejected or accepted and *merged* with the project. Often, changes to the proposed new code will be advised or required by the reviewer, providing an excellent learning opportunity for new developers. Having a first PR accepted is a major rite of passage, as is a first code review, while the satisfaction of knowing your code commands use within a large system like Python is something I can only guess at right now. Emotion is involved, legitimately. The problem is lack of modulation of that emotion.

Few programmers are better qualified to discuss the open-source process than Brett Cannon. The Canadian has been a Python core developer for two decades and currently oversees the Python extension for Microsoft's VS Code editor, an example of a new kind of "corporate open-source" project model—an innovation I might view with cynicism if VS Code didn't end up answering all my code editing prayers down the line. Over ninety minutes of measured but intense conversation Cannon describes everything Nicholas did and more. Often, when overt aggression is exhausted in an exchange, passive aggressiveness and guilt trippery steps into the void, perhaps with claims that the phasing out of an old, rarely used feature is a denial of future users' rights, likely to send them wailing to JavaScript. Or that turning down a particular PR amounts to trampling its advocate's right of participation in open source, even their First Amendment right to freedom of speech. Cannon had come to our conversation from just such a charge.

Friction is elevated by core developers' responsibility to take a holistic view of the code base. A perfectly workable new feature that benefits a few people may appear to its proponent to have no downside, but it will need to be maintained for the next ten-plus years. Even a harmless-seeming change can break one or more preexisting features or connections, forcing them to be revisited. And of course every online tutorial and millions of books in bookshops have just been made obsolete. Meanwhile the language is bloating and losing focus, is getting harder to learn, understand and manage. A change I and a few buddies might like to see comes at substantial unseen cost to someone like Brett Cannon and Python's millions of users.

Most Pythonistas understand this. When I ask Cannon what proportion don't, he says it's not the proportion that matters, but absolute numbers: a single abrasive interaction can destroy equilibrium for an entire day. He also says he thinks fewer than ten positive exchanges to each negative one is psychologically unsustainable—and that the ratio of good to bad is lower than that. He reached a point in 2016 where he, too, had to step back to seek help. And Cannon, a thoughtful man whose undergraduate degree is in philosophy, was most distressed by a single heartbreaking realization: he thought he was becoming a worse person, more impatient and angry and bitter. More like his assailants, in other words. The problem, he adds, exists only online. At PyCon even people coming to him with complaints or criticism do so with respect, prefacing their conversation with the appropriate *thank you for your service* tributes. Like most open-source core developers, Cannon gives his time to Python for free. Thanks is the least he and the many thousands like him deserve. For a significant minority of coders, the big-picture concerns he and other open-source maintainers live with appear to have no purchase at all. A curiosity of this situation is that no one I talk to believes the civility crisis in code has grown worse in response to the wider society. I am starting to suspect these parallel agitations are related, but in a more complex, symbiotic way than I can yet put my finger on.

🐞 🐞 🐞

A flash of further insight into the root of incivility comes from another source.

Coder *id* appears in rawest form on the website *Hacker News*. With a claimed five million users, 10 percent in the Bay Area and a third in Europe, HN triggers strong feelings internally and externally. Mature programmers revile the site, but a convocation of this size is hard to dismiss. Like it or not, *Hacker News* reverberates throughout the industry. This is where "well actuallies" banned by The Recurse Center go to get it on.

When the *New Yorker* comes calling on HN in 2020, a female MIT researcher whose software helped capture the first photo of a black hole has just made the front page. Writer Anna Wiener notes a typical response among *Hacker News* users: rather than congratulate one of their own on a significant achievement, they crawl through her code on GitHub to belittle her contribution.

"The site's now characteristic tone of performative erudition—hyperrational, dispassionate, contrarian, authoritative—often masks a deeper recklessness," Wiener observes. "Ill-advised citations proliferate; thought experiments abound; humane arguments are dismissed as emotional or irrational. Logic, applied narrowly, is used to justify broad moral positions. The most admired arguments are made with data, but the origins, veracity and malleability of these data tend to be ancillary concerns."

Please reread and remember the above passage: we will return to it.

The most startling part of Wiener's piece involves a moving portrait of the moderators charged to defend what tiny loop of generosity survives on *Hacker News*, fighting a 4chan-style slide into an even lower circle of Hell. Daniel Gackle and Scott Bell turn out to be uncommon people. With spooky perfection, they met in Calgary at a user group for Lisp, a now rarely used but much-admired language that retains the capacity to make even hardbitten coders smile. The stream of hostility—of bilious indignation, smugness and spite the pair confront every day—can only be met with a reciprocal sum of self-knowledge, as they try to model good behavior in the face of relentless aggressive pushback. Proposed software solutions to their problem have seldom made an impact, they maintain. What does work, if too slowly, is painstaking engagement with individuals. To this end the moderators lay their psyches on the line every day.

"In terms of the psychological experience of doing this job, all of your buttons are being pressed on a regular basis," Gackle confesses to Wiener. "The sheer quantity of it is so overwhelming that one does have a depressive reaction, a hopeless reaction to it at times."

Hacker News is ubiquitous and extreme enough to attract parody from appalled senior coders. The satiric website *N-Gate*, named after the NAND logic gate that outputs *False* if both inputs are *True* (droll slogan "We can't both be right") refers to HN users in the plural as "Hackernews." A representative item, headed "MIT to no longer consider SAT subject tests in admissions decisions," hints at the quality of debate on the site:

> *MIT makes extremely minor change to their application process... Hackernews is extremely concerned that colleges are deprioritizing arbitrary standardized testing, since those were the only things Hackernews was ever good at. Other Hackernews point out that these tests tend to return better results for rich people who can afford shitloads of preparatory work, and that this might be an unjust practice that needlessly excludes people without access to the preparatory training. Mostly, says Hackernews, those people can fuck off. If these kids are so smart, why are they poor?*

When Anna Wiener contacts *N-Gate*'s proprietor, she finds an engineer who grew up in Palo Alto in the heart of Silicon Valley but works in rarefied high-performance computing up the coast in the Pacific Northwest. He insists on anonymity but makes the seriousness of his satire plain by email.

"Almost every post deals with the same topics," he writes,

> *These are people who spend their lives trying to identify all the ways they can extract money from others without quite going to jail. They are people who are convinced that they are too special for rules, and too smart for education. They don't regard themselves as inhabiting the world the way other people do; they're secret royalty, detached from society's expectations and unfailingly outraged when faced with normal consequences for bad decisions. Society, and*

> *especially economics, is a logic puzzle where you just have to find the right set of loopholes to win the game. Rules are made to be slipped past, never stopping to consider why someone might have made those rules to start with. Silicon Valley has an ethics problem, and 'Hacker' 'News' is where it's easiest to see.*

Four things strike me about this assessment. First, it could be a specific description of the planet's most famous "hacker," Mark "they trust me, dumb fucks" Zuckerberg. Second, most of what's being said would equally apply to the finance industry and elements of the business and political classes. Third, key traits of the personality being described align neatly with those of a narcissist or psychopath. And fourth, if we didn't know before, we certainly know now that there is a logical but baneful irony encoded in the human condition: that certainty accrues disproportionately to the most ignorant and people who consider themselves of superior intelligence are invariably, almost by definition, *not* ("Real knowledge is to know the extent of one's ignorance," said Confucius). In this light, it should be no surprise that, toward the end of Wiener's *New Yorker* piece, Daniel Gackle, who by any definition of the word "intelligent" *is*, floats a disarming truth. One of his and Bell's most effective tools in pacifying Hackernews, he explains, involves the simple trick of trying to engage an aggressor's intellectual curiosity.

"It might seem like 'intellectual curiosity' isn't primarily an ethical concern," he reflects. "But it actually turns out that those things are deeply related. There's something about the way that curiosity works: it needs a kind of gentleness."

Please remember this going forward, too.

I don't know how much use I'm going to be when our second Py-Week game-building competition comes around, but the timing and tenor of our collaboration couldn't be better. I am happy to be working with Nicholas on what I can see is a release for him after the tensions around Mu. Better still, his idea for the project is an inspired one. Let's go old skool, he says, and create a *MOO*. Through a series of coincidences, I know what this means—and am instantly enthralled.

The term "MOO" is geek slang for a form of MUD, itself an acronym for "multi-user dungeon," the whimsical name given to a genus of text-based online games that attracted cult followings before sound and video came to the net, before most citizens had much idea of what the net was or would do. Of the few people who heard of MOOs, many (including me) were drawn by a 1998 book called *My Tiny Life*, in which author Julian Dibbell described an event that struck most readers as incongruous back then: a series of crimes, including a *rape* no less, within a previously flourishing online community called LambdaMOO.

LambdaMOO was a typical MOO, essentially a database describing a magical mansion in which users could adopt a persona of their choice and proceed to interact with others, moving through the house and even building their own rooms, chambers, tools and anything else that could be described in words. The server was in Palo Alto, but a several-thousand-strong community that lived in the house was dispersed and could have been anywhere. The point was precisely that you didn't know who or where anyone was in their alternate physical form. LambdaMOO amounted to an interactive, collectively written fantasy story you could enter and leave at will, and the community grew close. "Virtual" was about to become a buzzword, but hadn't yet. Most of us had scant idea what it implied.

The crimes against this community were committed by an evil clown named Mr. Bungle. No one was physically hurt in the "real" world. Like everything on LambdaMOO, Bungle's trespasses were virtual. And yet these events *felt* traumatic within a previously harmonious confederation, the more because no one knew who was bringing this discord or why. In 1998, most readers puzzled that events in cyberspace could have genuine emotional mass and even be visceral for a group of people who mostly didn't know each other IRL, that is, In Real Life. Now we can see this for what it was, a shot across the bows of our virtualizing world that almost no one noticed.

To the surprise of us both, LambdaMOO is still going. What's more, a LambdaMOO Programmer's Manual survives online as a template for other MOOs. I sneak an early peek and belatedly realize that when you boil MOOs down, they are nothing more nor less than primitive social networks that appeared three full decades before the term "social network" came into being.

The theme of this PyGame competition will be "six." At length we discuss what the conceit of our MOO will be, settling on Nicholas's relatively straightforward idea of creating a literary virtual world, with six starter chambers in which players converse thematically—starter possibilities being *Limerick Lounge*, *Shakespeare Salon* or *Tolkien Tearoom*. More important than these will be the provision of tools allowing players to create themed spaces of their own imagining. Scores could be awarded by fellow players for style and extent of contribution, in the manner of PyWeek itself.

So, we're going to try our hands at building a MOO. As a technical challenge this is fascinating. How does such a network operate?

Over beers one balmy night I seek advice on our PyWeek project from my neighbor, Sagar. On an evening such as this, as the retreating sun casts a poppy-gold glow over the burnished slopes of our canyon, you understand how easy it is for Californians, especially older and wealthier beneficiaries of the property and tech lotteries, to think of themselves as somehow chosen. A phrase I hear often is "Why would you live anywhere else?" Or as Governor Jerry Brown once put it, "Where ya gonna go?" We are about to be reminded that nature has a quiver full of hubris-piercing answers to this kind of question. But for now, watching young redwoods sway in a Pacific breeze as though dancing toward us, while local coyotes howl and turkey vultures glide from the sky like royalty to roost at the edge of our woods, it's hard not to feel lucky to be here.

One of the least complicated pleasures of moving to California has been getting to know Sagar. In his early sixties now, he works as a senior quality engineer at one of the more visible and socially conscious software giants in San Francisco—which description could tempt a person to make easy assumptions about him. This would be a mistake. We first bonded over a shock shared enthusiasm for an obscure late sixties to early seventies English musical outgrowth called the Canterbury Scene, but the fascination of Sagar for me is in watching someone who takes such joy in building things, to whom engineering in the broadest sense is like breathing, fixing things elemental. Occasionally I get involved in these projects and while we work together I learn the story of how he arrived at software, which is by some measures eccentric, others typical. Into the bargain comes insight into how the tech cliché we all laugh at, "making the world a better place," came into being—and perhaps also how it could be reclaimed.

The first inkling that Sagar might benefit from some piece of neural equipment my brain never had or sold to the devil at the crossroads for a packet of Cheetos came early, when our households decided to grow some vegetables on a small patch at the back of his and his wife Tara's yard. When mystery vermin started eating the new plants, Sagar announced he had a plan. I went round one afternoon to help action it.

My impulse would have been to spray something ugly at the bastards. Sagar had other ideas. I arrived to find he had been to the hardware store and bought all the materials we would need to construct large, portable, chickenwire-wrapped cages to cover the plants, removable for the purpose of weeding. Over a pleasurable couple of afternoons, we made them while discussing a private software project of his aimed at fixing America. I wasn't sure software could do this, but soon saw I was missing the point. More than anyone I had ever met, the point for Sagar was the process. He loved making stuff. When he wasn't making stuff, he was thinking about making stuff. When a fence blew over, he was out there digging up and repouring concrete to make new foundations. Another time I went round to help move some earth, to find he had constructed an *actual crane* for the purpose out of hooks, pulleys, hinges, swivels, two-by-fours and a bucket. Oh, and the earth was needed for a giant model railway planned to run the entire perimeter of the property, with viaducts, tunnels, towns, industrial clusters and turnarounds, which saw Tara—who may have preferred a rebuild of their crumbling deck—rolling her eyes with mildly exasperated good humor. That was Sagar: What could you do? He builds stuff. Try to stop him and he turns back into a frog.

Was it nature or nurture? Who can know? Sagar's father was an engineer, working with real trains for the giant Bechtel Corporation, mostly abroad after America made its tragic pact with Detroit and backed the car over public transportation (still, the Cheetos were good). Yet Sagar's grandmother told a story of how once when he was a very small boy, she made the mistake of switching off her vacuum cleaner and leaving the room for a few minutes, returning to find it strewn across the floor in a forest of pieces. Not knowing what to say, she barked "put that back together!" So, he did. Of course, she didn't call him "Sagar" back then, but we'll come to that.

Even in the Bay Area, computing wasn't on many kids' radar in those days and knowledge of what these machines did came indirectly. Sagar's mom worked at the fabled Bell Labs for a time but saw no route to advancement as a woman and left. His high school administration subscribed to a computer scheduling service that supplied crazy-cool timetable printouts on the first day of the academic year, like something from *2001: A Space Odyssey*, but brought chaos in a way that still amuses.

"The schedules would be *so* wrong! Every time there would be three times too many students trying to jam into one room, with none in another and the PA announcing that all students assigned to *this* class go to *here* while we try to figure something out."

So, the first thing Sagar knew about computers was that they could not schedule students for class ("I know now that this is a hard problem," he grins). Later, as a shy, nerdy physics major at Cornell he had access to a computer, an IBM System/360 for which you wrote programs on punch cards in a "truly weird" language called PL/1 (a link to the original MIT hackers). You couldn't touch anything: you submitted your program to the "high priests" of the machine, who returned it the next day "with twelve giant sheets of printout telling you why your program didn't work." Computing had all the fun of a tax audit in those days.

To his family's despair Sagar dropped out of Cornell. Some part of him wasn't being nourished. Like millions of disenchanted young Americans at that time, he moved to the country and joined a commune, the eighty-strong Twin Oaks clan in central Virginia, based on ideas the behavioral psychologist B. F. Skinner expounded in his book *Walden Two*. And this was a good experience, felt like a real education, shared with smart people questing for a more sustainable path than the America of Watergate and Vietnam and the gas crisis seemed to promise. Disillusioned with psychedelics and macro politics, many countercultural refugees had turned to computers as a vector for hope. Sagar arrived at Twin Oaks just as the communards accessed a groundbreaking Apple I to help with accounts.

He learned BASIC and immediately wondered what freaky stuff you could make the machine do. With a computer-literate coadventurer he threw himself into *Project Leap*, an ambitious plan to rearrange accommodations across Twin Oaks's five residential buildings, allowing members to live in closest proximity to their closest friends. The pair conducted a survey of preferences and decided to write a program to run as a simulation, with people as particles that had attraction and repulsion values in relation to others. Clever. Would people clump together in groups that matched the buildings when the program ran? *No!* They careened to the edge of the screen. Undeterred, they tweaked numbers and tried again, to see half the group fly edgeward while the rest chased each other around to infinity . . . and Sagar thinks they might have got there in the end had the communards not lost patience with the machine and found their own solution. "I still think I'd like to try and solve that problem one day," he muses, pointing out that the presence of the computer had encouraged people to consider the problem, conduct the poll and find an analog workaround, so its net impact was positive—an early lesson that solving a problem for the computer is not always the same as solving it for people.

Sagar stayed at Twin Oaks for four fruitful years, then landed a coding apprenticeship in Upstate New York, working for a female Jamaican coder-mentor-boss ("try to imagine that now") in a new database program-cum-high-level-language called dBASE II. But by this time something even more inviting had appeared on his horizon. Twin Oaks was a secular community, but spiritual seekers turned up from time to time, among them followers of the Indian guru Osho. Previously enlightenment-skeptic, Sagar thought Osho's teachings looked like the real deal. And the deeper he dug into them, the better he felt.

If the name "Osho" rings a toebell, it will be from the hit Netflix TV series *Wild Wild Country*, about the giant Rajneeshpuram commune he and his "sannyasin" followers ran on an Oregon ranch from 1981 to 1985. What the documentarians never made clear, Sagar points out, is that sannyasins *paid* to be on the ranch, which meant that he, having no money, couldn't go. Until one day a German friend called with news that nonpaying American sannyasins could come to the ranch without charge, that for a limited time enlightenment in Oregon would be free. Enthusing "That's my price!" he handed a day's notice to his bemused but supportive boss and set off for the West Coast.

Almost immediately he sensed a dissonance. The problem was not Osho's collection of ninety Rolls Royces, which Sagar understood to be contributions from supporters in lieu of money (for which the guru—being a guest wherever he went—had no use). But where Twin Oaks had been a democratically run cooperative, Rajneeshpuram was highly stratified and authoritarian, with the guru's Machiavellian lieutenant Sheela calling the shots. New arrivals were not told they had been summoned to vote down townsfolk in an upcoming local election, a maneuver that culminated in Sheela poisoning the citizenry in an unsuccessful effort to keep them from the polls. Vote-rigging plan foiled, unmoneyed sannyasins were of no further use to leadership, so Sagar and his cohort suffered the indignity of being sent to Portland. The future programmer had been on the ranch less than two months and, like many former sannyasins, blames Osho's naïveté for Sheela's misdeeds. Still, he made lifelong friends and met Tara as a result. Osho's ideas, never clearly articulated in *Wild Wild Country*, still resonate. Just as importantly, a few of Sagar's sannyasin acquaintances had both money *and* computers.

Sagar started reading *Byte* magazine. He taught himself C and assembly and experimented with games until he'd learned all he could from them. By 1984 the first IBM PCs and Apple Macs had appeared, and he knew this was the path for him. He was way behind the experts but didn't care. Humankind was embarking on a profound journey and he had a seat on the bus. It was thrilling.

He found another dBASE II job, at a Seattle auto garage with a sideline in Vespa scooters, earning peanuts for the chance to learn. And his employer got what he paid for, Sagar confesses with a chuckle, "because I did some *BAD* things!" To mitigate the tedium of stepping through slow-drawing menu screens on a 48K machine, for instance, he located the physical address in memory of individual keys to create his own bespoke shortcuts. Ingenious! Until a new version of dBASE arrived and went berserk. By which time, fortunately or otherwise (depending on whether you were Sagar or his boss), the budding software wizard had chased a girlfriend to California.

San Francisco furnished a first grown-up programming job, at a company writing complex software to help maximize the longevity of underground mines. The firm could afford one very fast multi-user computer, which everyone had to log in and out of.

"But get this," Sagar laughs again. "It used *Fortran*. In 1987! And Fortran is a godawful language. Any language where you have to care about line numbers is *bad*. What line you're on is part of the language."

Line numbers mattered because Fortran made integral use of the *Go To* command (e.g., "go to line xx"). In this way you could jump around the program at will, but once line 16 consisted of "Go To line 47," what happened if you added a new line between 16 and 47? You had to remember to go back and change line 16 to "Go To line 48." And if you didn't, the program would crash. Dijkstra wrote about Go To in terms mostly associated with crimes against humanity. Ultimately Go To would be replaced by functions, little coded routines—sometimes still called "subroutines"—that were allocated names and could be called from anywhere within a program, removing the need to park the code for a specific task (say, *print*) on a specific series of lines and jump to those lines when their functionality was required. (When we want to print something in Python, we simply write `"print('whatever')"` wherever we need it to happen, and the interpreter finds the code relating to the print function and runs it.)

"It took a lot of convincing to get the assembly language types to realize Go To—writing line numbers into a program as part of the program—was bad. These are still all over even high-level languages once you get down to the assembly code level, but they're abstracted away at the higher level."

Sagar felt he needed to understand how the hardware worked, so took night classes and bagged his first big company job in hardware at the Bay Area animation giant Pixar. Soon he was also coding there, before moving at the invitation of a former colleague to Sonic Solutions, spinoff of the local Marin County powerhouse

Lucasfilm, keeper of the *Star Wars* franchise. He joined a startup that got bought by one of the big players, then in 2006 was recruited to a small business software shop that grew into one of the Bay Area's most visible behemoths, Salesforce, whose founder is known locally for—among other things—supporting a local big business tax to deal with the homeless problem when few of his counterparts would. Here Sagar remains. Almost everyone will have encountered the software he helps build, and as a quality engineer he spends a lot of time thinking about how company code behaves in the wild. Temperamentally inclined to see the best in things, he loves his job and finds it deeply absorbing. Most of the engineers I get to know do: the degree to which programmers love to program is astounding. When Sagar's company decided he should pick up Python in the interests of machine learning, he did it in a week. If I didn't like him so much, I would feel obliged to hate him.

Sagar has clear advice for Nicholas and my PyWeek MOO challenge. TDD. Test-Driven Development. One of the most counterintuitive ideas ever introduced into programming, closely associated with the groovily named *Extreme Programming* and related *Agile* software development paradigms. Concepts owing their existences in great part to a man named Kent Beck, without whose decades-spanning contributions modern computing would look very different.

As much a view of the universe as an approach to writing code, TDD—and its cardinal idea *unit testing*—was formulated in opposition to the "waterfall" philosophy that holds sway in most creative fields, including conventional programming, whereby an artifact (in this case a computer program) cycles through progressively more granular stages of development and refinement until finally considered "finished" and ready for release, with testing an afterthought. And with traditional media this made sense. I can't come back and rewrite this sentence because I've decided it rather got away from me, wasting space on redundant words like "rather" and containing unnecessary detail about Kent Beck being a polished guitar player and us sharing a love of goats: we're just stuck with it. But in code we can effect change and refinement *after release*, while our programs are running—in fact we must, because the only universal truth about code is that There Will Be Bugs. Beck was not alone in saying, "Why not release the minimum spec that works, then refine in situ?" Which looks like a recipe for chaos. Except the developer had another brilliant idea to prevent the perpetual hostaging of *Homo sapiens* to half-finished code. He wrote the first *unit testing* framework, SUnit, for the now seldom seen but wildly influential language *Smalltalk*.

Used judiciously, unit testing and its extension test-driven development flipped coding on its head in a mind-bending way. Instead of organizing functional code into a program that seems to work, then adding any testing code afterward or—more usually—not at all, Beck advised writing tests *first*. At every stage we ask ourselves, "How would we know our program is running as intended right now?" Say player KingGeorgeV has just entered our Limerick Lounge. He should have permission to be there: if he does not, something is wrong in the code. So we write a simple Boolean test. *KingGeorgeV has permission to be in the room*, True or False? Without functional code to satisfy our test, it should fail by returning false. Now we write a few lines of functional code, always the minimum necessary to make the test return True. We might add another few lines telling the program what to do should our test fail in the field. After which we ask ourselves "Now what needs to happen next?" Then we write a test for that and repeat.

Ideally, no functional code will ever be written until a Boolean True/False test exists for it to pass.

We would expect code written with such rigor to be reliable. Also long. But according to Sagar, reliability is only one of TDD's benefits. For one, we proceed in small, incremental steps and avoid the near-universal feeling of being overwhelmed by a new project. TDD also ensures a bug is easy to trace and will not crash the whole program, because instructions on how to handle a failed test are written in. More important still is the way unit testing compels developers to think their code through in detail beforehand. In committing to write tests first, one asks and answers practical questions about what the code should do, and how, questions that might otherwise be ignored until it's too late.

"It turns out that you can view the tests as kind of a definition of what the program should be," Sagar explains. "There's this flip that can happen, where you suddenly realize the tests are as important or more important than the code that fulfills those tests. Because they're like a road map. The tests can be so good, and so full, that they primarily define what the program should do. The program itself becomes secondary."

Perhaps most importantly, TDD helps manage change as features are added, bugs fixed, elements rewritten. Sagar mostly codes in his favored Java these days, but there is a Python unit testing framework called PyTest. He suggests we use it. I find myself wondering if unit testing could be adapted to writing and other pursuits too.

With delectable serendipity, Nicholas has already decided to use PyTest for what he christens our "TextSmith" MOO platform. In the event all I can do is watch, read, learn and cheerlead as he flies into a coding fury, devising a Boolean test consisting of some condition that must be met, then a snippet of code to satisfy the test, then another test and more code, and another, and another—all while waiting in hallways for kids to finish band practice; at home as the family sleeps; in the morning when everyone's been packed off to school; between his own coding day job and jazz piano lessons taken up in the wake of the Mu ordeal . . . And if all this writing-testing, writing-testing sounds like Hell's own sinbin, what I see is Nicholas having the time of his life, summoning something beautiful as if conjuring it from air, much as I see Sagar doing on a serial basis next door. I hope one day to find a project to inspire me in such a way, though any such prospect feels remote right now. I also get the first inkling that there may be some larger yearning behind Nicholas's mania of invention, a longing expressed in his admission that "The thing I miss about MOOs are the narratives and creativity. Facebook, Twitter, et al. constrain our self-expression to a dribble of easy-to-mine data points. I wonder what a MOO-ish social network would look like for today's world."

I don't know it yet, but Nicholas is far from the only coder to be asking this question in 2019, and I wonder if he might be looking for a way to return to that earlier Eden, at least for a time. An appealing proposition, to be sure.

A revelation for me is the extent to which a classic Moo's logic and language reflects and illuminates that of the machine. At its heart is an engine called a "parser," which takes input from a user and translates it into "action" within the game. The parser does this by recognizing the elements of a standardized sentence containing a verb, direct object, preposition and indirect object. The parser also accepts a stripped-down construction of just verb plus direct object. Articles like "a" and "the" are not used. If I am playing as the knight "Dirk" and input "slay dragon with frisbee," the parser processes each word based on syntactic position: it will search its database for objects labeled "dragon" and "frisbee," and a verb given as "slay." Assuming these words are found, the parser changes the attributes attached to each object according to instructions written into the verb "slay." From this point on, no one else can slay the dragon. As we have seen before, the parser doesn't "understand" the words as we do—it recognizes relationships between them and can work only with objects, verbs and prepositions that have been written into it. Input of "discombobulate dragon" is sure to draw an error message meaning "verb not found," where "eat dragon" is probably fine, "smoke dragon" okay in California.

How does the parser make and keep track of such a web of relationships and changes? The way a computer does everything. Each object in the game is allocated an identifying number and *points to* other numbers that represent other objects to which it maintains connection. Behind the curtains, a bunch of numbers are being pointed at each other and connected, disconnected, reconnected according to instruction. The numbers are then converted back to words upon which humans have bestowed meaning. Without our imaginative projection these numbers remain just numbers, at once banal and weirdly divine, the dark energy of twenty-first-century life. Again we see exactly what Paul Ford meant back in New York when he declared that "Computers are basically faking everything . . . they don't know how to do anything except add numbers very fast. They're frauds. *Calculators.*"

Objects are brought into being and removed from play with simple functions called *create ()* and *recycle ()*. So here we have a low-bandwidth virtual reality environment that lays bare the mechanisms of our digital world. Facebook may be more complex, but its underpinnings are the same.

🐞 🐞 🐞

At the end of PyWeek the genius of Nicholas's MOO choice becomes clear. Because there are no Python libraries pertaining to MOOs, all code has had to be written from scratch around a scaffolding of tests. Where *Traffic Flowmageddon*, with its liberal use of prewritten modules, ran to five hundred lines, TextSmith accrues a thousand—a third given to unit tests. But when the comments start appearing, one complaint we don't hear is "This doesn't work/is full of bugs." TextSmith amounts to a conceptual rack for players to hang their own hats on—and that was all it needed to be.

Nicholas is expecting gruff feedback, but it doesn't come. One competitor named DROID doubts that a text-based game is "appropriate" to PyWeek and grumps about having had "no fun reading all the text," but others very clearly have had fun and love the retro cool of the concept. A typical comment begins, "This has great potential. The object attributes system is neat and the choice of font and markdown really invites creating a rich literary world." This commenter follows others in offering sharp observations on what the game lacks and how it could be improved, but these faults are excused in the face of PyWeek's time constraints. Another player enthuses that "I had quite some fun poking around this one. I hope we get more text games." Yet another notes that while TextSmith's commands were a little awkward at first, they "worked well when I got the hang of it," adding that "the possibilities are endless and you could easily spend hours making or interacting with the game." An innovative facility for players to embed audio and video links in the literary salons, which turns TextSmith into a "multimedia MUD," attracts broad admiration. "Folks have been building fun stuff!" Nicholas enthuses, and by now I can see how much the sensitive, empathic, inventive man I've come to know needed this. If the open-source trolls could see the impact they had, would they reflect on what they'd done? Or would it encourage them further? It suddenly seems important to find out. In the meantime, when votes are in at the end of the assessment week, we are stunned. TextSmith has come fourth in the team competition and sixth overall, an amazing result for such a singular effort.

Our revisitation of MOOs teaches me a less obvious and more immediately practical lesson. When it comes to assessing PyWeek entries in order to vote, *I can't*. In an unwelcome echo of the previous PyWeek, TextSmith is one of the few games I can make work.

The problem is *dependencies*: outside applications or modules from a library, *imported* into our program to perform specialized tasks. In *Traffic Flowmageddon* we used modules (of prewritten code) from the PyGameZero library to generate zombies, buses, taxis and so on. If someone had already written code to perform these tasks, why waste time writing our own? Instead, we imported the module at the beginning of our code ("import pgzero"), co-opting functionality while keeping it firmly in its black box so we didn't have to think about it. This is a wonderful thing. But it does mean that Python must be able to find PyGameZero in my local file system to run *Traffic Flowmageddon*. In code, as in life, a gnarly tradeoff accompanies convenience.

My system does work with PyGameZero. But competition games frequently use complicated gaming libraries that don't play well with the MacOS operating system I still use (most pros commit to the open-source Linux). Other more exotic and arcane dependencies abound and many clash with some unknown element among the haphazard mess of lost apps and vestigial software littering my hard drive like limpet mines. In nearly all cases, workarounds and fixes exist, but these mean diving deep into my hard drive, messing with files whose names alone give me the bends; which I start to think of as "gray boxes" containing code I can access but would be too terrified to touch. Again I am reminded that syntactic mastery is important, but what keeps pros up at night are points of confluence, where codebases (and the human animal spirits behind them) come together like phantom players in an orchestra. Like every wannabe coder before me, I am going to have to find a way to get comfortable with this truth, or at least intelligently uncomfortable, because—as Ben Franklin would have agreed—the only certainties in twenty-first-century life are death and taxes and doubtful dependencies.

There is a deeper and more universal revelation, courtesy not of any of the tools Nicholas and I used, but of a prime champion of those tools: Kent Beck himself. As Sagar explained the principles of unit testing to me, Beck had just done his FAANG time at Facebook, leaving after a seven-year stint. Whether a blog dated February 24, 2018 goes any way to explaining his departure is unclear. What he writes is arresting.

"Twenty-five years into my career and I finally figured out my mission," he begins, going on to enumerate the challenges presented by a world not built for people like him, who "can't read social situations" and are exhausted by group interaction, who are "wired for symbol manipulation and not social interaction." He appears to be describing a personal point on the vast autism spectrum, though he doesn't say this. Either way, his challenge is finding ways to feel safe in situations he can't easily interpret. He continues:

> *Sometimes when I feel awkward or unsafe it's because I'm being awkward or I'm in an unsafe situation. Sometimes, though, I'm actually safe and I just feel unsafe. I want to address both situations. To address feeling awkward because I'm actually awkward I can learn social skills. I'll never be as good at them as someone who is better wired for social interaction and/or has practiced them more than I have. That's okay. I don't have to be great at socializing, I just need to suck less. To address feeling unsafe in spite of being safe I can learn to better think my way through social situations. I can insert logic between stimulus and my response. I can also remove myself from situations when I've hit my limits. I often do this after talks. Ten minutes alone lets me deflate myself and return my energy levels to normal.*

Part of Kent Beck's newly discovered "mission" is to arrive at a position where his feelings about a particular social situation reflect actuality, allowing him to feel "safe" when he is and unsafe when he is not. Another is helping to explain people like him to people unlike him. To "help geeks feel safe" in the world. Weird as Beck is by his own account, he is part of the human race and only "whole" when connected. "I could probably figure out how to be a pretty happy hermit," he admits, "but I'd be missing out."

The rest of us would miss out more. I look up some of Beck's talks and find a charismatic speaker from whom original ideas skip like stones on a lake. He's funny, too. Beck's self-described "weird" mind leads him into intellectual spaces others might not see: his contributions to programming have been real and sustained. Most recently, he claims to have been hired to help fix Facebook, to have spent five years assessing the problem then been fired for offering solutions. He acknowledges that he can be abrasive and offend others without meaning to or even understanding he has, or why.

But there's more. Based on recent insight, Beck ends his "Mission" blog with a striking disclosure. All his innovations in the ostensibly hyperrational field of programming, he says, have been driven and directed by *emotional* impulses. His first unit test framework aimed to "transmute worry into tests that replicate confidence" and foster feelings of security. His test-driven development aspired to "break overwhelming problems into a stream of test-code-design-test-code-design" that would protect against big-picture overload, or paralysis by context. My guess is that the only unusual feature of the subjective hierarchy Beck describes is that, with the help of a new wife, he has been self-aware enough to observe it. This is a profound insight with substantial implications, the extent of which only become clear to me as I head deeper into Silicon Valley.

CHAPTER

11

The Gun on the Mantelpiece

Truth is much too complicated to allow anything but approximations.

—John von Neumann

Shortly before the start of PyCon 2019, a bombshell. Guido van Rossum, Python's creator, touchstone, court of appeal; the closest thing a computing community could have to a spiritual leader, is retiring as BDFL, *Benevolent Dictator for Life*, at the age of sixty-three. A wave of love and appreciation sweeps through community noticeboards, promising to lend a special intensity to the conference. For most of Python's life, contentious changes or disagreements among core developers could always be settled by Guido, with his decisions commanding automatic respect. How to fill the void left by such a totemic figure? Discussions have already begun. The sense I get is that, to many, the Creator's retirement feels not just like the end of an era, but the loss of a kind of innocence.

PyCon is in Cleveland again. For all my continued frustrations with the limited time and energy I have for learning, my favorite talk will be an ostensibly dry technical one, delivered to a packed mid-size theater by an experienced engineer named Jack Diederich. "How to Write a Function" should be dull as dishrags but isn't. If all art aspires to the condition of music, all Python aspires to the condition of tedium, until the striving itself becomes perversely compelling. Diederich captures this dynamic with style.

Based on advice he gives in code reviews, the engineer begins by asking the existential question, "What is a function for?" It is perfectly possible to write programs without functions, as a series of statements describing data and operations to be performed on it. An example: our town gift shop sells Shakespearian insult kits. The web is full of free DIY ones. In most cases these consist of three lists of slurs assembled from the Bard's plays by Samaritans-in-reverse, two given to adjectives and one to nouns. A user chooses one word from each list to form a withering Elizabethan calumny. I use mine daily and have long dreamed of automating the process. Would this require a function?

After thinking and sketching awhile I figure the obvious way to start is by declaring our three lists and assigning each to a separate variable. We then write code to draw one word from each variable/list and combine them into our Shakespearian dis. Python has a built-in "random" module full of *methods* to facilitate randomization (technically pseudo-randomization, because true randomization is impossible in the deterministic fief of computing). One such method, "choice," returns a random element from any collection passed to it—in this case our lists. As with all modules and their methods, "choice" needs to be *imported* into our program before it can be *called* into action. Our setup could look like this:

```
1 from random import choice
2
3 adjA = [
4     "mammering",
5     "saucy",
6     "beslubbering",
7     "spleeny",
```

```
8       "yeasty",
9       "mewling"
10 ]
11 adjB = [
12      "rump-fed",
13      "plume-plucked",
14      "onioneyed",
15      "clay-brained",
16      "beetle-headed",
17 ]
18 noun = [
19      "bugbear",
20      "flap-dragon",
21      "hedge-pig",
22      "joithead",
23      "lewdster",
24      "moldwarp",
25 ]
```

There are many ways to mix randomly chosen words into the combinations we require. To start with, the clearest and most obvious (to me) looks like this:

```
1 from random import choice
2
3 adjA = [
4       "mammering",
5       "saucy",
6       "beslubbering",
7       "spleeny",
8       "yeasty",
9       "mewling"
```

```
10 ]
11 adjB = [
12       "rump-fed",
13       "plume-plucked",
14       "onioneyed",
15       "clay-brained",
16       "beetle-headed",
17 ]
18 noun = [
19       "bugbear",
20       "flap-dragon",
21       "hedgepig",
22       "joithead",
23       "lewdster",
24       "moldwarp",
25 ]
26 word1 = choice(adjA)
27 word2 = choice(adjB)
28 word3 = choice(noun)
29 insult = f"{word1} {word2} {word3}"
30 print(insult)
```

The magic happens in lines 26–9, where each list is in turn *passed* to the "choice" method, which selects one item at random and assigns it to a unique variable (word1, word2, word3). In line 29 the values contained in our variables (in this case insulting words) must be combined into a three-word *string* of characters, ready to be printed to our display. As we know, strings in Python are delineated by double or single quotation marks. But standard strings are *literals*, meaning anything within the quote marks will be interpreted exactly as it appears. And we don't want to print the name of the variable, we want the value it contains. The solution is to precede our string with an "f" to create an *f-string*, a relatively recent Python innovation I learned about on one of the wilder nights at a PyCon bar. In an f-string, anything contained within curly brackets will not be interpreted literally, but in terms of the value it represents. This value can be in the form of a variable or—brilliantly—an *expression* that requires calculation. For instance, f"{2+2+2}" will be interpreted within an f-string as "6." Line 29 therefore forms the three words stored in the variables word1, word2 and word3 into a string, then assigns this string to a new variable, "insult," to be passed to the familiar "print" function in line 30. And presto! Sir/Madam, thou art a:

```
1 => spleeny plume-plucked flap-dragon
2 => yeasty beetle-headed bugbear
3 => mewling rump-fed moldwarp
```

Except that, because f-strings will handle expressions within their curly brackets, we can effectively do all our work within one f-string. If the following causes undue brain-ache, skip to the first paragraph of the next section. Otherwise try parsing the logic in line 26 of this stripped-down insult generator, from innermost brackets outward, passing lists to method to f-string to print function:

```
1 from random import choice
2
3 adjA = [
4     "mammering",
5     "saucy",
6     "beslubbering",
7     "spleeny",
8     "yeasty",
```

```
9       "mewling"
10 ]
11 adjB = [
12      "rump-fed",
13      "plume-plucked",
14      "onioneyed",
15      "clay-brained",
16      "beetle-headed",
17 ]
18 noun = [
19      "bugbear",
20      "flap-dragon",
21      "hedge-pig",
22      "joithead",
23      "lewdster",
24      "moldwarp",
25 ]
26 print(f"{choice(adjA)} {choice(adjB)} {choice(noun)}!")
```

```
1 => mammering clay-brained joithead!
2 => beslubbering onion-eyed lewdster!
3 => saucy plume-plucked hedge-pig!
```

A new Bardish insult is generated on each run of the program, with no function necessary. But there are problems. Simplicity may be a virtue in general, but here we have no way to prevent the inevitable duplicates. And wouldn't it be comely to allow the continuous generation of insults until a user either chose to quit or became aware they had no friends left? Yes, but now we have a much more complex set of problems, making a function seem wise. I think and sketch and google some more, and at length end up with a thirty-line program incorporating a *for loop*, a *while loop*, two *if statements* and five separate variables, marshaled by not one but two functions, the first called from *inside* the second. With so much going on I figure this is pretty clever, not least because to my utter surprise it works. I go to bed feeling pleased with myself . . .

And wake to the castigation "*What was I thinking?*" My original specification for the program included functionality that I thought would require two functions. When I rationalized the spec, I kept the dual-function architecture, mostly because I was intrigued by the technique and found it fun. But given the new spec it was also insane. From my earliest *Python Crash Course* learning I recall a built-in function called "pop," which allows one item to be "popped" (removed) from a list and worked with. I now see that, given this capability, I can use a for loop to generate not one insult at a time, but a list of, say, one hundred insults. If I set a condition that only unique insults are to be placed in the list, with duplications discarded, then the list will contain no doubles. Another built-in function, called "input," allows a user to supply input or make choices while a program is running: I can use this to offer users a choice to keep popping and printing insults from the list, one after the other, until the list is exhausted or they opt to quit. Because the list contains no doubles, neither will the output. In this way one function (plus one if statement and two variables) from version one become redundant, shrinking the program from thirty lines to twenty-two. Here's what the Shakespearean Insult Engine v2.0 looks like (again, explanation to follow):

```
1  from random import choice
2
3  adjA = [
4      "paunchy",
```

```
5       "reeky",
6       "lumpish",
7       "rank",
8       "goatish",
9       "surly"
10 ]
11 adjB = [
12      "tickle-brained",
13      "weather-beaten",
14      "sheep-biting",
15      "swag-bellied"
16 ]
17 noun = [
18      "codpiece",
19      "puttock",
20      "minnow",
21      "maggot-pie",
22      "pie-face",
23      "varlot"
24 ]
25 def insult_engine():
26     bardish_insult_bank = []
27     for i in range(100):
28         insult_i = f"{choice(adjA)} {choice(adjB)}
           {choice(noun)}!"
29         if insult_i not in bardish_insult_bank:
30             bardish_insult_bank.append(insult_i)
31     next_action = input("Generate a Bardish
       insult? (y/n): ")
32     while next_action == "y":
33         print(f"\n{bardish_insult_bank.pop(0)}")
34         next_action = input("\nPrithy another? (y/n): ")
```

```
35      else:
36          return f"\nFarewell, thou {bardish_insult_bank[0]}\n"
37
38 insult_engine()
```

```
1 => rank swag-bellied codpiece!
2 => goatish tickle-brained puttock!
3 => reeky sheep-biting pie-face!
```

For development purposes our "adjA" and "noun" lists will contain six words and "adjB" four, affording 144 potential combinations. More typical forty-word Shakespearian insult kit lists would enable 64,000 unique possibilities. Line 25 defines the function we are writing, while line 26 declares an empty list and assigns it to a variable given the name bardish_insult_bank, ready to receive generated insults. The fun starts in line 27, where we set a for loop and use Python's "range" built-in function to set a number of iterations for it to carry out. In human terms this line translates as "for each number (i) in the range 0–100, do the following"—or more straightforwardly, "Do the following one hundred times." We can choose any number we like and, as always, anything indented immediately under the loop's declaration in line 27 is part of the loop: this is what will be done one hundred times. Line 28 deploys the f-string formula described earlier to generate a random three-word insult and assign it to the variable "insult_i." Because we need each slur within our list to be unique, we set a *conditional* in the form of an if statement in line 29: if the fresh insult generated on each loop (insult_i) is *not* already in the bardish_insult_bank list, then (line 30) append it to the list.

Thus, on each iteration of the loop, the insult_i variable will be overwritten with a new insult, checked for duplication within the bardish_insult_bank list and either discarded or added. We get a clear view of Boolean logic in action here: the conditional provides a simple statement, "insult_i is not in bardish insult bank." If the interpreter finds this statement to be true, it carries out whatever instructions follow the colon. If the statement proves false, the interpreter ignores those same instructions—and in this case carries on looping until one hundred iterations have occurred. One hundred iterations should produce a list containing close to that number of items, allowing for discards, but if we increase the iterations to one thousand, say, it should contain all 144 possible combinations of the words in our adjA, adjB and noun collections. Crucially, each run of the program will present the available insults in a different order.

Now we have a list of 144 Shakespearian insults to play with. In line 31 we use Python's "input" built-in to offer users a choice of whether to receive an insult from our insult bank or quit the program. The way we do this is by passing input a message, in the form of a string, to present to the user. Input invites a response and records the user's keystrokes, which in this case we are storing in the variable "next_action." In lines 32–34 we now define a while loop based on the input we received. This says: "While the value contained in next_action equals 'y,' perform the tasks described in the indented lines immediately below." Line 33 makes use of the pop function I woke up from dreaming about. Computers always count from zero, not one, so "pop(0)" means pop the first item from the bardish_insult_bank list. By enclosing this in an f-string we can pass its *value*, rather than literal name, to the print function for display. "\n" simply instructs the interpreter to create a new line, like the return key on a keyboard.

In line 34, still within the loop, the user is asked for yes/no input again, and so long as the input is "y" our while loop will continue to cycle. Because the first insult is being removed from the list on each iteration, and because we took steps (in line 29) to prevent duplicates from entering the bardish_insult_bank list, every successive pop will present a different insult. Should the input function be instructed to assign anything other than "y" to the next_action variable (line 35), the while loop will break and (line 36) a final insult will be popped, inserted into a string and displayed, to read something like:

```
=> Farewell, thou paunchy weather-beaten
maggot-pie!
```

Finally, in line 38 we *call* the function, causing it to run.

Refinements could include displaying the insults in bold or color; providing options for users to save favorites or customize the adjA, adjB and noun lists; introducing error messages to warn of incorrect input; displaying the content of bardish_insult_bank or placing limits on how frequently individual words can appear in the output.

🐞 🐞 🐞

Fine: We know this by now. But what are we really doing when we write a function? What should we be *trying* to do? Like all the best communicators, Diederich breaks his issue down to a point where it feels like a succession of statements of the obvious—though bitter experience insists it is not. A function, he says, consists of just three things:

(a) An input, which could be data we supply ourselves in the form of a dictionary or list, as per our insult engine, or from something non-local to us, such as a web page, spreadsheet, remote database or application programming interface (API, used to access data from a remote server)

(b) A transformation of the input into something we want (this is where we add value)
(c) An output, where we *return* what we've done

"Really," Diederich says, "a function should be unexciting."

I will learn that in the real world of programming, gathering the required input (step a) is often the most delicate and even dangerous, because it may involve interfacing with other machines and programs (and therefore, indirectly, *people*) over which you have limited control.

"In your real day job, the first thing you have to figure out is how do you get your information. So you have to figure out what you have and what you want. And to make it easier for the reader in the middle, throw out what you don't need," Diederich says.

For the first time I notice the way experienced Pythonistas address "the reader" with the same concern authors do, more so in some cases. By the time we get to step b, the transformative core of the function/algorithm, "the reader should be bored," Diederich says, to a rumble of laughter. Output should be in whatever form is most useful to the user of our program, whether it be a new list or dictionary, a stream of data or a new file of the same data in a different format.

None of what Diederich says sounds novel—until he explains why he's taking the trouble to talk about functions, because these straightforward-seeming "objects" turn out to be the digital equivalent of potato salad, with theories about content and construction running the gamut from redundant (three types of potato) to cheeky (capers); alarming (kale); or felonious (raisins). Neither is this always the fault of coders, because some of the most pointlessly prescriptive stipulations, according to Diederich, come from senior code reviewers such as himself. Why, he wonders, would anyone insist that a function should be no more than two lines long before returning an output? But they do. In the early to mid-seventies, when C was introduced and even sophisticated computers had just 48K memories, such an injunction might have made sense, being easy on the compiler that converted human code to machine code. But not in the twenty-first century, where memory is cheap and plentiful. Another common edict is that all functions should be as compressed as humanly possible, drawing gasps of admiration that so few words and symbols can accomplish so much work. Diederich shows three versions of the same short function, each getting tighter and cuter and more likely to win admiring gasps, before revealing that his team had gone not with the super-svelte editor-genic third, which introduced a couple of extra confusions for the reader—"which at this point included the guy who wrote the code"—but with the slightly longer but crystal clear second. At the wackiest end of the spectrum is an insistence that conditionals in the form of "if" statements, one of programming's algorithmic workhorses, should be abhorred, necessitating all manner of gratuitous contortions.

To make a broader point, Diederich draws on the metaphor of "the gun on the mantelpiece," invoking the theatrical law that if there's a gun on the mantelpiece in act one, it needs to go off in act three, building suspense in an audience that understands the convention and can "enjoy the journey as the moment approaches." He pauses before drawing raucous laughter with:

"This is absolutely *not* what you want in code. If there's a gun, you want it to go off now—we don't want suspense, wondering what will happen, in code."

The point is to introduce new elements and concepts only when they are needed. So, if I need to instantiate an empty list for potato salad ingredients,

```
pot_salad_bits = []
```

I should do it immediately before the code written to fill it. My instinct up to now has been to create my lists and dictionaries up front so they're in place when I need them, which would make sense in the physical world, where the word "object" is not a metaphor and things have to be manufactured. But not here, where I can make whatever I want, whenever I want, with a few keystrokes.

"Help the reader by using as few concepts as possible at any time. Make the job easier, everyone's happier," he advises, before concluding to the kind of applause more usually found at a Taylor Swift show:

"So, in summary, when you look at a function, think input-transform-output. Use sensible names and avoid unnecessary work. I mean, don't clever up the place."

If being back in Cleveland feels good, returning to PyCon as a sophomore feels better. I may remain the worst coder in town this week, but at least I'm no longer the newest. Am I alone in feeling a bittersweet edge to my joy at being in this manifestly sane and bighearted place, with Ernest Durbin III (later to become known as Ee) once again urging us to review the Code of Conduct and remember the "Pac-Man rule" of always leaving room for one more to join when chatting in the corridor, while noting the three gender-neutral bathrooms and cautioning that the pretty peppermint green 2019 souvenir t-shirts are sheerer than usual, so make sure we're comfortable in them before going out and maybe don't choose this year to burn our bras? *Sigh.* Perhaps thoughtfulness and uncomplicated goodwill should always have seemed this precious. Given the now perpetual rancor outside our little Pythonic bubble, I know I'm not the only one thinking, "Can't we all just stay and live here, like this, at PyCon?"

Since my foray into open source, I have a keen understanding that no language is an island and there are things to worry about here, too. As last year, one of the largest patches of promotional real estate in the main hall belongs to Facebook. Again festooned with signs expressing hunger for Python, the Facebook stand crawls with visitors at a time when the company's eagerness to profit from misinformation is proving as constant and unconditional as Fox News's—and with even more pernicious consequences. By year's end the stream of Facebook scandals will be so constant that it resolves into a permanent state, a dangerous new normal. I visit the stand myself and browse one of the glossy recruitment brochures that detail expanding areas of the business and what kind of coders are sought to drive them. Out of thirty-four departments, including "search," "global consistency" and "applied machine learning," only one mentions security and is concerned with the company's protection of itself from us rather than vice versa. In this context, the "detection" department looks like the grudging afterthought it is.

I corner one of the young staff members and ask if would-be recruits are mentioning the whole "undermining of democracy tending to outright promotion of fascism" thing as a possible disincentive to joining the team at 1 Hacker Way? He looks a little pained, but not defensive. "No," he replies with a sheepish shake of the head. "To be honest, I was expecting a bit of it, but no one's mentioned it at all." Later in a corridor I bump into Eric Matthes, author of my beloved *Python Crash Course*, who I met and became friends with at PyCon last year. He looks pensive as he tells me he did the same thing at the Facebook stand and was disturbed to get the same result. Even in a community set up to be as intentional as this one, he says, it's as if pro coders have a tacit agreement not to fret the ramifications of their work and to stick to technical discussion, denuded of responsibility or consequence.

Nicholas joins us and suggests a degree of self-selection, with more conscious Pythonistas avoiding the Facebook stand altogether. And he may be right. All the same, I have seen this before, at an SF Python meetup in the gorgeous deco Yelp building in downtown San Francisco, where I watched a senior engineer from the company being treated like royalty, with not a mention of the nefarious activities her work enables. This was a star Pythonista with a dream resume; she could have worked anywhere. Later still, at PyCon, a Facebook "Evangelist"—a job title that once seemed sweet but now looks chilling on the business cards of Big Tech recruiters—will introduce a keynote event with the words "I am extremely proud to work at Facebook," again without a word of challenge or visible wincing from anyone.

On the first night, a treat for Nicholas and me. Dinner with Guido. Assuming we can get out of the building before eateries have closed, past the gauntlet of handshakes and selfie requests, tongue-tied introductions and expressions of love for the language the Dutchman gifted the world. Half an hour after meeting in the corridor, we finally make it outside to a sidewalk and are en route to a serviceable sports bar when I find a young man who turns out to be from Chennai in India walking beside me. He catches my eye.

"Is that *Him*?" he whispers.

I nod *yes* and the young Pythonista introduces himself to Guido, saying "I'm so excited to meet you. And now I don't know what to say."

"That's OK," Guido smiles. "You can just say that."

Respectful interruptions and gifts of beer continue through our meal, always met with good grace and humility, never impatience. With his gentle demeanor and playful eyes, my first impression of Guido van Rossum is that he would be a very hard person to dislike, and I will leave agreeing with Nicholas that Van Rossum's personality is a key ingredient of the language and community he founded. I also agree with PSF Chair Naomi Ceder, a classicist before she was a coder, when she tells me one of his most distinctive gifts to codecraft has been to prioritize aesthetics, crediting it with an innate ethical dimension. In this sense I've come to think of Python as both modernist and identifiably Dutch in sensibility. Guido laughs when I say this, then cocks his head to one side, thinking, as we agree to meet back home in Silicon Valley to discuss further. By then he will have embarked on his post-BDFL life of biking and hiking "until I get bored." Before we break up, he does confirm his decision to pull back from Python as having been made the morning after he approved the controversial Python Enhancement Proposal (or PEP) admitting the Walrus Operator into the language. Like Nicholas, Brett Cannon, Russell Keith-Magee and so many others, he was wearied by animus. The sort of people you want to see at the center of a community, the sort a community *needs* to stay healthy, are precisely the ones being driven out. When I ask Naomi Ceder how the open-source ethic came about, she describes putting the same question to a group of computing eminences at a conference and being told it was in opposition to the siloed, compartmentalized, closed mores of the Cold War and Vietnam, that early computerists "wanted community and cooperation for the world *they* were building."

I'm not the only newbie to marvel at what those early coders built. One of this year's keynotes comes from a coding alum of San Quentin State Prison's thriving Last Mile program, which teaches inmates to code. In this instance, the speaker was drawn less to code itself than the cooperative open-source ethic. "To me," he says, "this was the very definition of love." Open source changed him by example. And I know exactly what he means: I have begun to feel the same. Watching the best being driven out by the worst is heart-wrenching—doubly so because this year it seems to be happening everywhere, within and without code, as if by contagion.

At the urging of friends I put a card on the board to run an Open Space session discussing these issues in the context of the book I'm writing. To my pleasure more than thirty people turn up. I tell them about my background and struggles with the task at hand; of my despair at JavaScript, misadventures with machine code and still deepening affection for the Python community and open-source ideal. I describe my wonder at the way they make this uncanny microcosmos of "bits" do anything at all, let alone some of the rhapsodic things I've witnessed at PyCon. Also my fear that, as elsewhere in society, constructive voices are being drowned out by a minority of antisocial trolls, while too many coders are still too tightly focused on their own code at the expense of what it does when it strikes the world. I'm also interested in their view of their own image outside the community, which coalesces around guilty enjoyment of their wizardly perception undercut by concern that the stereotype of the eccentric, highly specialized genius—invariably pictured with fingers flying across a keyboard at an impossible rate—may deter viewers from imagining themselves as coders.

When it comes to monitoring the impact of one's own code, generational differences appear. An older participant points out that you don't always know the setting in which your code will appear, and often never see the overall program into which it will slot like Lego. "You don't really feel like you've got power," he adds. "I always assume if I don't do it someone else will." Another senior engineer describes refusing to contribute a line of code he feared would encourage the misuse of data. A pair who work at Comcast confess to knowing the negative reputation of their company and being affected by it. "Our engineers have developed some really good open-source tools, but they never get any traction outside the company because no one trusts us," one says with sincere-seeming regret. In timeworn tradition, young coders in the room are fiercer and more idealistic, more likely to insist that "We have to be responsible for what we write." A majority surprise me by welcoming the European Union's recent General Data Protection Regulation (GDPR) privacy legislation, grateful for the leverage it gives against nefarious code, arming engineers with the only argument many twenty-first-century management teams appear to understand: that breaking an actual law and being prosecuted might be bad for business. External pressure makes it easier to rebuff management teams drawn to the shadows. By this account non-coders have a powerful role to play in tech, turn out to be part of the development environment. Perhaps placing responsibility on individual coders' shoulders is both unfair and impractical. They need our help.

For all the concerns around programming and culture wars, PyCon abounds with examples of code at its magical best, code that illuminates the world and makes possible what otherwise wouldn't be. At a full lunch table (seven men, three women), one woman works as a contractor for NASA, writing code for the Hubble Space Telescope, a busy job since the US military donated two surplus lenses ("Their technology is way ahead of everyone else's," she smiles. "But of course their lenses point down instead of up"). By coincidence a second guest at our table works at NASA's fabled Goddard Space Flight Center, writing software to track orbital debris, much of which stems from a Chinese anti-satellite missile test in 2007. The software he develops can track fist-sized objects and warn of collisions five days in advance, and is improving all the time. Together we attend an Open Space session called "Python in Space," whose title suggests a low-budget sequel to *Snakes on a Plane*, but which ends up being serious fun. Most of the forty or so coders present came to Python—often after Java—via science, and describe projects ranging from creating fleets of tiny "cubesats" to producing images from radio telescopy and using machine learning to search vast expanses for signs of light bending around objects invisible to the eye. We hear with a little spark of pride that the recently captured sound of two black holes colliding was produced using Python, perhaps with help from the new Astropy library.

Like everyone present, I walk away from "Python in Space" filled with the joy of life. Where else would I meet these people? At a space conference, I guess. But then who would teach me how to make a neural network track the behavior of crows in my yard? With no small shock I realize it's getting hard to imagine my life without these exotic meetups. Who could predict that, for the next three agonizing years, that is exactly what all of us here will be forced to do?

"But you're already infected!"

Sagar's response when I told him I was having my brain scanned a few months into my coding education. He was only half-joking. In Magdeburg Professor Siegmund pointed out that anything new will change your brain, but Sagar shares my intuition that learning to code changes it in distinctive ways, meaning it alters who you are and how you receive the world. Asked if he thinks anyone can learn to do it well, he says *yes*, if they really want to.

"But it's gonna change you and you have to be open to that. Some people won't be, they'll resist, and those people will find it hard."

That was me, I said; maybe still is.

My scan results contain some surprises. No dead badgers are found, figuratively speaking; no neural hornets' nests. And once catastrophic shocks have been ruled out, seeing images of my own brain, lit like a nebula and inspectable in three dimensions, seems fantastical. Is that me in there? Scientists are starting to say *no*—that our brains, minds, selves can only be understood with reference to the gut; nervous system; internet; human organism at large. But the brain alone contains mystery enough. Have you ever wondered what a daydream or memory consists of? How do you "see" something that plainly isn't there? Or by what outlandish mechanism we form a complex stream of thought into intelligible sentences with no apparent process of composition? A computer can be known at any given moment: it consists of a sequence of discrete "states." But a brain confounds the notion of stasis, is so fluid that the sole way to fix and define its state would be to kill it. All we know for sure is that the human brain is the most mind-blowing thing we have seen in the universe so far, the only thing outside the nether reaches of the internet capable of being blown by itself . . . or of grasping the irony that a half-formed dream of exceeding itself with binary abstractions—code—could conceivably lead to its own destruction. We are indeed remarkable organisms.

The lone notable physical feature of my brain is its large volume. Norman Pietek is quick to point out with a chuckle that size has no bearing on function and the only consequence of outsize dimension is difficulty fitting into a scan frame. In parsing code snippets, however, there are meaningful departures from what the team has previously seen. To recap: Janet Siegmund and her team have scanned dozens of coders calculating the output of Java code snippets. Until I turned up, study subjects shared a high degree of commonality, being mostly male computer scientist undergrads at a German university. Analysis of the results turned heads because code comprehension in these highly proficient programmers appeared highly lateralized; localized to regions associated with syntactic language processing, working memory, problem-solving and close attention—a pattern common to natural language comprehension. Regions of the right hemisphere associated with math and higher logic were little engaged. Taken as a whole, these scans seemed to suggest that our brains process code similarly to natural languages—that the designation "programming *language*" is not fanciful. Given that music is broadly associated with the right hemisphere in right-handed people, this account would posit any preponderance of musicians among programmers as coincidental. Throughout, Janet Siegmund reminds me of the limits of fMRI technology and the simplifications involved in interpreting scans, that the whole brain is in use all the time. When the background activity is filtered out, however, the above is what we see.

The most obvious contrast between my response to code snippets and that of previous subjects is also the most interesting. Where previous results indicated a high degree of lateralization to the left, mine do not: my scans show abundant activity in both hemispheres. Discussions with the team yield a favored explanation for this more diffuse activity: that my novice status forces my brain to work harder as it gropes for meaning—to be less efficient than that of an expert with thousands of hours of practice. Perhaps this narrowing of activity expresses what expertise *is*. Dr. Siegmund predicts my second scan will look more like the others. I'll be curious to see.

What am I learning about the existence of a "coding mind"? Fitful as my learning has been, the microcosmos has proven more fascinating than I could ever have guessed—and yet it still feels alien and uncomfortable. Had the circumstances of my life been different, could I have been a really good coder? Or is it true that great ones are born, are different from the start? The joker in the 10x pack is a continued, pop culture endorsed presumption that preternatural ability corresponds to a specific personality type. In his "My Personal Mission" blog, Kent Beck spoke of being "wired for symbol manipulation and not social interaction." A gruff and unwitting insensibility to others' emotional states is as indivisible from 10X mythos as raw skill. When I think back to my first PyCon, I am a little shocked to detect my supposition that the remote guests at my PyLadies auction table, and the man who sat next to me at a Great Lakes Science Center dinner the next night without ever taking his headphones off, were code savants, fleeting visitors from a parallel dimension where even the fundamental structure of thought confounds my own. But why, on no evidence at all? I realize the time has come to seek out some acknowledged masters and see how different they really are.

CHAPTER

12

Code Rush

Everybody is a genius. But if you judge a fish by its ability to climb a tree, it will live its whole life believing that it is stupid.

—Albert Einstein

My favorite times at PyCon are the communal breaks and meals, where tables fill fast and computerists submit to chance encounter. One lunchtime I found myself in a group containing a mix of male and female commercial coders from around Europe, skewing east, plus one Argentine astronomer and an Indian woman physicist who used Python to analyze data. Amid this international collection was a soft-spoken American who piqued my interest partly for the reason I piqued recruiters' at corporate stands—he was my age—and partly for how he was dressed, sporting just enough Code Freak markers to suggest a pedigree. Also, where most of our PyCon badges bore first and second names, his showed only the one he'd offered on introduction: "Wolf."

The conversation was lively until most of the table rose to attend talks, one of which, "Randomness in Python: Creating Chaos in an Ordered Machine," had been on my schedule, too, before curiosity got the better of me and I stayed to talk with Wolf. As the huge dining space emptied and we found ourselves alone save a few stragglers and staff, I felt free to pose the question I couldn't before. Was "Wolf" his full name?

"Yes, it is," he replied.

He invited me to imagine names as gifts from our parents; how we might be impelled to refuse that gift if we didn't much like those parents. His had given him *Scott Collins*, but by the age of sixteen he knew he wanted another identity, even if finding the right one and going to the immense trouble of legal reassignment had to wait till his forties. Asked if a single moniker causes problems, he affirmed that it does—*lots*, from a teenage son assumed high when stopped by police and asked for his father's details, to a wife given the third degree upon trying to deposit a check, to a suspicion that job applications wind up in piles reserved for people who can't be arsed to fill them out right. The judge who granted the change warned him, he confessed, and she was far from wrong. Yet not for a minute did he regret his mononymous state, because not only is "mononymy" a beautiful word, but he was "reminded every day that I am my own person and I formed my own value system, and I am what I am because of me, not because of how I grew up."

Somehow I knew Wolf was a special coder, while understanding he would never say as much. One clue was the unaffectedness of his view that "programming in Python is like coming home to a puppy," which made me laugh so hard a bypasser thought I was choking. I told him my puppy seemed to spend a lot of time biting chairs and peeing on the floor just then, and asked if it would get easier. Not easier, he smiled, because the better you get, the more difficult the problems you're set. But you learn to enjoy the problems and the challenge of acquiring the skills you need to solve them. Even the documentation gets interesting to read, he said, which was unimaginable as infinity to me at that time.

I liked Wolf. He spoke interestingly, was also curious and a careful listener. After forty cordial minutes we parted for separate talks, but he stayed on my mind through the afternoon, like a haunting, without my quite knowing why. Only later did I find out.

❋ ❋ ❋

The 10X legend produces strong reactions in my orbit. Within Python it has limited currency. Gifted programmers abound within a community of this size, but a gift that put its owner on a different plane from others would be seen as having little value; as a solipsistic throwback to an age before the microcosmos dilated into the complex, shifting continuum we know today—in which it is rare for coders to create anything of worth on their own. When I spend a day with Tom Natt, chief engineer at the UK Government Digital Service, recognized as the best in the world at public software (and the model for its US counterpart), I mention the "10X" idea and he shoots back, "Oh, we wouldn't want that here." UKGDS is and needs to be a team, he explains, whose members rely on each other and collaborate selflessly as a unit. Where would the lone wolf 10Xer fit into this model?

Paul Ford urges caution in relation to 10X lore when I ask him about it, while admitting he does have a friend who fits the description, an American who lives in Berlin and looks like a Navy SEAL, who climbs mountains and is as far from the code nerd stereotype as it gets, even if he *is* a touch eccentric. And yet there is no denying that "the sparks in his brain jump some pretty wide gaps." Closer to home, Nicholas believes in 10Xers to about the degree he believes the British royal family are reptiles, possibly a little less, and I sense mild disapproval at my stated desire to meet some of the supposed chosen. All the same, when we compare notes in a bar the night after I've met Wolf, his response sounds a lot like awe.

"Ah!" he says. "Do you know who that is? There's a documentary called *Code Rush*—that'll tell you about Wolf—watch it when you can."

Now I do.

Code Rush is anchored in the first dot-com boom, a carnival of idealism and avarice that introduced the web to the public in 1995—before detonating it over Wall Street five years later. We now call this first iteration of the web "Web 1.0," and little of it was spared when the crash came in March 2000. Even so, those five years provided as phantasmagorical a ride for a handful of the switched-on young as anything seen in the 1960s. If you could code, even just HTML, you had the status of a magus.

The film tells a poignant story from that era. Up to 1995 the internet was a utopian subculture: most Americans had never heard of it. But in August that year, the runaway IPO of a first commercial web browser, Netscape, changed cyberspace forever. Overnight everyone wanted a piece of the action, despite almost no one understanding it. Unfortunately for the small Netscape team, the ensuing "net rush" caught the attention of Microsoft and its rival browser, Internet Explorer . . . a rivalry only one side was ever going to survive.

We arrive as the Netscape team accepts the inequity of their situation and makes the bold decision to go open source. But the open-source process has costs: since choosing this righteous path the code base has ballooned thirtyfold to over two million lines. As launch date looms the share price is fragile, having already fallen 500 percent from its dizzy post-IPO peak, and with programmers still chasing bugs all over the shop. No individual can hold such a giant codebase in their head, so problems often become apparent only when tranches of code are assembled at the end of development, the worst possible time to find them. Fear is undercut by gallows humor. The task looks impossible. And at the center of it all—knock me down with a feather—is *Wolf*.

The fascination of *Code Rush* is its fly-on-the-wall window on how coders saw themselves in the first phase of a revolution that would draw them to the center of the culture—which is as a maverick band of micronauts voyaging boldly into a new frontier. If the *Starship Enterprise* ran on donuts and Coke, this would have been its crew.

The handful of Netscape programmers all grew up around code. One of their moms was a programmer and director at Sun Microsystems, where Java was developed. All tell a coding origin story I hear over and again, about the first time they were able to make something happen on the screen, describing an almost drug-like high often felt as a sense of control or even *power*. I'd wondered at the significance of the tender age at which this usually happens, when feelings of power are scarce—until I experienced it for myself, when the subjective thrill took me aback. John Lennon once said the message of rock and roll was *be here now*, and in its quirky, maddening way the same may be true of coding for those who discover it young.

Wolf, for his part, knew he wanted to code from his first flirtation with the machine. By the age of fourteen he was earning fifty bucks an hour setting up and fixing computers for local businesses (another story common to top programmers). One of his team likens coding to a youth subculture, akin to being a mod or goth or punk, because you need to be able to throw yourself into it unreservedly even while knowing it won't last.

"There's no stability here," he points out. "So, it's a very kind of weird . . . *irony* that the very people who are inventing the future can't see their own future."

You realize these young mid-nineties coders take perverse pride in an image of themselves as driven freaks. A male manager offers nothing by way of contradiction: "These guys will just keep going till they're done. They don't need food, they don't need sleep. They *maybe* need pay, but that's all."

While a female counterpart smiles as she likens managing coders to herding cats. "You want them to be cats. You like cats and need them to be cats. But they don't all want to go in the same direction."

The film's climax comes a day before the software is due to ship, with the share price wobbling and investors needing the merest excuse to bail. A piece of Apple software indispensable to Netscape's operation has not been cleared for use, placing the project at risk of ignominious collapse. Fresh from pulling an all-nighter to fix the fiendliest remaining bug, an exhausted Wolf is forced to spend a second night writing his own scratch version of the missing Apple application, a seemingly impossible feat that he nonetheless finishes at 5:40 A.M., allowing Netscape to ship. Needless to say, permission from Apple arrives shortly thereafter, followed by an offer from the major web portal AOL to buy the company. Netscapers are in for a payday.

Happy Ending! One would think. But we see a somber side to the mania too. The filmmakers claim there were seven million coders in the United States when *Code Rush* was shot, adding that it was the highest-paid profession in the country on aggregate. Yet these figures mask the fact that a substantial portion of most tech workers' pay during the dot-com boom took the form of stock options, ensuring they only did well while stocks were flying (and most stocks would be worthless after the Great Dot-com Crash of 2000.) One of the younger Netscape coders understands this only too well.

"Stock options are a con, it's a carrot they dangle; it's like if you give up your one and only youth maybe someday you'll be rich. I know so many people who gambled on the startup lottery and got nothing. It has to be taken as a stupid tax. But I happened to win that particular lottery."

He has, too. News of AOL's bid rockets Netscape stock to $170 per share. Even so, two militant open-sourcenik engineers leave. Another retires, burned out beyond repair, with his wife ruing the toll startup coding life has exacted on her marriage, windfall notwithstanding. She advises others to think before leaping: "You need to count the cost, because you can't ever retrieve the time that's lost."

Wolf sounds a more chilling—and prophetic—note:

> *In the Valley, if you've stayed someplace longer than about three years people wonder what's going on, why you can't get another job—what's wrong? If you're a programmer you expect to change jobs every two years or so. It's like ants, worker ants. They send a group out to do something and as that group approaches the task they're there to do, some ants leave, more ants come on, and by the time it gets to the target it could be a totally different set of ants. I think as we distribute the work more and more in the information age, it'll be more like that for everybody.*

Is he saying what I think he's saying? That society will evolve to mirror his code? The open-source Mozilla organization was established to maintain the project. Enthusiasts for the Firefox browser are using the inheritor of Wolf's work.

When I asked the publisher and Valley maven Tim O'Reilly whether he thinks coders are born or made, he paused for a moment, then said, "I guess I would say both those things are true. Anyone can be good at it. You can learn to write serviceable code, just like you can learn to write serviceable prose. And then some people astonish you."

Who had astonished him? I wondered aloud, expecting a weighing of options that didn't come.

"Andy Hertzfeld," he said simply.

Which is how I come to be standing on the stoop of a handsome nineteenth-century Craftsman-style house in the leafy Professorville Historic District of Palo Alto, curious as to what kind of person I'll find behind the carved wood door.

Andy Hertzfeld. I already know the name. Played by Michael Stuhlbarg in Danny Boyle's Oscar-nominated biopic of Steve Jobs; primary creator of the first Apple Mac's game-changing operating system and graphical user interface (or *GUI*, pronounced "goo-ee"), arguably the most radical features of that deeply radical machine. And yet this is the least intriguing of Hertzfeld's achievements, because in three subsequent startups he pioneered the ideas of cloud and mobile computing fully fifteen years ahead of time. He did his obligatory tour in Mountain View at Google but, being a lover of music and literature—of Pynchon, DeLillo, Nabokov, Roth, Joyce, Dylan—who openly approaches coding in the way of an artist, his eight years there were frustrating.

"Programming is the only job I can think of where I get to be both an engineer and an artist," he has said. "There's an incredible, rigorous, technical element to it, which I like because you have to do very precise thinking. On the other hand, it has a wildly creative side where the boundaries of imagination are the only real limitation."

Struggling "to exercise my creativity in a way that gives me joy" at Google, he left.

Do these biographical details align with the 10X construct? In themselves, maybe not. But as Hertzfeld seats me in a beautiful period living room full of dark wood paneling and books, a documentary on the most storied of his post-Apple startups is about to stream and leave no doubt that if "10Xer" means anything at all, the man perched on a chair in front of me is one.

General Magic tells the story of Hertzfeld's second and most tantalizing startup. The introductory banner "Failure isn't the end, it's the beginning" hints at the narrative arc. But it's also true. By the heartrending climax an early nineties team of General Magic engineers drawn heavily from Hertzfeld's Mac group had established the basis of mobile and cloud computing, USB connectors and the smartphone; had conceived and built advanced touch screens and invented emoticons. Mobile phones were still new—and *analog!*—meaning GM had to persuade AT&T to create a discrete network for their digital technology. When AT&T execs came to talk, their limo was too big to enter the car park. This is what the young startup team was up against. I watched *General Magic* with a mildly reluctant Jan, and by the end she was confounded to find herself in tears of frustration and sympathy. Never mind that from the technical staff Megan Smith went on to become America's chief technology officer under President Obama, Tony Fadell invented the iPod leading to the iPhone and Andy Rubin cofounded and led the Android open-source mobile operating system project.

"So, between Tony and Andy [Rubin] that's 98 percent of the world's smartphones," says Marc Porat, the CEO romantic who lost his company and marriage in the same year and took a long time to recover.

GM alums fanned out across the industry and had hands in too many innovations to list here. But the point is this: in a workplace replete with brilliance, Andy Hertzfeld stood out, was regarded with a kind of awe by his peers. Indications of why abound in the film.

"We would say, 'Hey wouldn't it be great if we could do *that*!' laughs iPod inventor Fadell, "and Andy would stay up all night coding that thing that we just came up with that evening . . . he would stay up all night and bring in the fully working thing the next morning!"

One day Hertzfeld wondered how to express emotions on a mobile network, so went away and came back with animated emoticons. When the team agreed to make a major change to their browser, Hertzfeld completely rewrote the system over the weekend ("So we came back and it's all there . . . that shouldn't have been possible for one person to do!"). And so on. Unlike his former boss Jobs, everyone seems to have liked Hertzfeld, too. In common with Apple cofounder Steve Wozniak, he considered *Jobs* a great movie but diverged by questioning its accuracy, engaging in an extended philosophical debate with screenwriter Aaron Sorkin about the ethics of storytelling—a discussion I would like to have been a fly on the wall for.

The maestro is retired now and only programs for pleasure. Even at his relatively modest age he claims to be no longer able to embrace the prodigious amount of information needed to visualize a complex program's shifting web of relationships; to hold them in his head the way an elite chess player or physicist does and he once did. Curiously, he's one of the only programmers I meet who sees the microcosmos as I've come to see it. Where most emphasize the prosaic foundations of computing—the mechanical march of bits within a known, definable environment—

Hertzfeld's company was called General Magic in all seriousness. When he told Apple he didn't want a business card because the job title options were too boring to commit to print, Steve Jobs told him to make up his own. Famously he chose *Software Wizard* "because you couldn't tell where it fit in the corporate hierarchy and it seemed a suitable metaphor to reflect the practical magic of software innovation." Always wearing a big, open smile in photos, this would have been an inspiring person to work for.

We talk. Unusually for the 1960s his high school had a computer terminal linked to a remote timeshare machine. The teacher in charge had taken a summer course and didn't know much, but even back then programming was something you mastered mostly through solitary exploration. So while other kids played sports or did drama, he learned BASIC and Fortran and was soon exhausting the school timeshare budget, having to invent projects to justify the expenditure. He wrote a computer dating program for his junior prom. Summer jobs involved computing for local businesses.

"I loved it," he says. "It was like, you build stuff—that was what I liked about it. You have an idea and then you realize the idea by applying logical thinking. And then you can watch the idea running."

He experimented with computer graphics, plots, sine functions; thought programming was fun and cool. And yet he saw no place in it for himself.

"I had a misimpression even of what computers were, you know. I thought they weren't for me because you had to wear a jacket and tie at a bank or maybe for the government. And that didn't align with my interests."

At Brown University he expected to major in math, a subject in which he showed precocious ability. But as he climbed into more advanced graduate theory in his sophomore year, math stopped being easy and fun. Missing the groundedness of code, he switched to computing, which then fell under the rubric of applied mathematics. Low-level work writing device drivers came naturally, but over time he was drawn progressively higher, to the most human facet of his discipline, *user interface*. Software developers rarely embrace these lowest- and highest-level pursuits together and Hertzfeld's talent for the latter seemed the more precious to him, so he focused on that. His boss Steve Jobs used to say that what thrilled him was "the intersection of the humanities and the sciences," and young Herzfelt felt the same.

Hertzfeld has a typically nuanced perspective on the 10X debate when I ask. He cites the work of Howard Gardner, the Harvard professor who postulates at least eight different types of intelligence, in suggesting that anyone might achieve 10X equivalence—or reciprocal incompetence—depending on the problem they're dealing with at a given moment.

"I would say it's not a quality of the person alone, but a particular programmer can resonate with a problem and they'll be fantastic at a given task, maybe ten times better than the average person—or even maybe the person sitting next to them—just because that problem resonates with their interests and thinking. So, people can be ten times more productive, but it's not just their characteristic, it has to do with the marriage of you with what you're doing. I just know from my own programming that sometimes I go ten times faster than other times. So am I '10X'? Is someone else? I don't know."

Coding today is fundamentally different from his Macintosh time, he says, not just because it tends to be done in teams.

"The big difference is Google and to a lesser extent Stack Overflow, in that when you're stuck you can tap into the collective wisdom of humankind. That's just a tremendous thing. These days, when I'm programming I spend maybe a third of the time googling solutions, or especially bugs, looking to see if other people have encountered them. And before open source we didn't have so many good programs available to learn from or apply to the task at hand. Remembering is still useful for going fast, but it's not the necessity it was."

Hertzfeld has played an outsize role in shrinking computers into the world's pockets, making software the interface not just between people and their machines, but between people and other people. I ask if the results have surprised him. There is a long silence while he thinks.

"Well, we had a dream when we were working on the Macintosh that someday everyone would have their own computer and love using it. And now . . . you know, I never would have thought then that you could overdo it. So, people are addicted to their phones and sometimes engage in unhealthy practices, ignoring the world around them. That bothers me. But the alternate facts and fake news thing, I don't blame software for that, I see that as a consequence of human nature. And the industry is trying to come up with tools to help people judge what is worthwhile. It'll get better with better tools, better techniques. The problem with a company like Facebook isn't technical. It has to do with values. I believe Mark Zuckerberg is insincere. His stated goals are not real: he's trying to maximize power and profit. The question is, is he willing to earn a little less money to make a better experience for everyone? And that's why I think he and Facebook are being misleading, just paying lip service."

If I want to learn how to help build those "better tools," he tells me, the best way is simply to write lots of code.

"That's how you learn. It's like riding a bicycle: You can't read a manual, right? You just have to do it."

Why am I surprised to learn that Andy Hertzfeld's favorite language these days is Python? I guess I thought it would be something hair-raising like C or Lisp. But there is a nice symmetry to the news as I thank him for his time and embark on the short drive north to Belmont, where the original Pythonista awaits.

Guido lives with his Texan wife and their son in a handsome modern house on a pleasant suburban street. There is no Lamborghini in the driveway; no trophy art on the walls; no photos of the celebrated programmer and languagemeister smiling with tech royalty. Furniture is stylish modern in the Danish-Dutch mode: clean lines, uncluttered, pleasing of form and generous in devotion to function. Even the soft furnishings whisper hints of Python.

Guido van Rossum was raised in Haarlem, the colorful Medieval city for which the more famous New York district—anglicized to Harlem—is named. His mother was a teacher prior to his and two younger siblings' arrival, while his father was an architect. By his own account he was exactly one of those smart, quiet kids many of us recall lurking at an amiable but mysterious tangent in school, huddled among the other smart, quiet kids with outlier hobbies like trainspotting or archaeology and a pronounced tendency to surprise former peers in later life by doing noteworthy things. In talks and writings Guido paints himself as an intensely nerdy child.

"Yes, I really was," he chuckles as we sit down with a cup of tea. "I don't think I was aware of just how nerdy I really was. Just as well, probably. When I look back I think, 'Gosh, I was so sort of *young* and *innocent* and *naive*.'"

By high school he was a keen student and hung with a group of kids who organized their own field trips to study nature, which remains an abiding interest. When I ask if he minded not being among the cool kids, he muses "a little bit," mostly because the boys in that group were more likely to have girlfriends. But overall the picture is of a model happy childhood. Like me he remembers a father forged before Ikeafication killed basic lifeskills, who was at home in books or at a drafting board and, when the family needed furniture or art for the walls, would *make* it, but who also had the people skills to communicate with contractors, teams of colleagues and difficult clients between amusing his kids with cartoons. Like me Van Rossum feels a little impoverished by comparison.

Guido also had a schoolboy's interest in electronics and circuitry, assembling primitive hardware until writing code to control the hardware began to look more fun. Like Andy Hertzfeld, he set out to study math as an undergraduate in Amsterdam before discovering a mainframe computer in the university basement. After graduation in 1982 he spent thirteen years at the renowned Centrum Wiskunde & Informatica (CWI), the Dutch national research institute and site of Dijkstra's important early work on algorithms, helping among other things to create a new programming language for use by nonexperts. Aspects of that language, ABC, would inform Python, not least the use of indentation rather than punctuation to delineate statements (as opposed to the more standard semicolons or brackets common to C-type syntax). ABC never took off, but from 1989 Guido set to combining its lessons with elements of other languages he admired, notably the "remarkably powerful and elegant" Modula-3 he encountered during a summer internship at the Digital Equipment Corporation (DEC) in Palo Alto.

To my mild surprise Guido and Dijkstra never met, though Dijkstra lore was ever present. He recounts with amusement how one renowned DEC theoretician created a font to mimic the great man's handwriting, allowing anything from manuscripts to memos about tidying the kitchen to be graced with the dramatic "EWD" imprimateur. Computational gravity drew both Dutchmen to America. From 1995 to 1998 Guido was given the opportunity to work on Python at the US National Institute of Standards and Technology, and from 1998 to 2000 at the Corporation for National Research Initiatives with additional funding from the Defense Advanced Research Projects Agency (DARPA). Following a long and successful technological tradition, in 2005 the state turned Guido's carefully incubated language over to the private sector when he went to Google, where he stayed for seven years before jumping to Dropbox. In contrast to Andy Hertzfeld, Guido was never drawn to the business side of his profession. Having to fire people upset him and even today he doesn't like delivering negative feedback. For him programming was never a procedural refuge from other people, and he professes to find meaning in it only as part of a community, taking pleasure in the humility this demands. In delivering a speech to mark the Dutch national holiday King's Day, he claimed to be "probably somewhere on the autism spectrum," bolstering my present impression of "the spectrum" as being broad enough to flirt with redundancy (a view that will later be confirmed by a specialist).

I find Guido in a relaxed mood even for someone whose base state is easygoing, far from bored with the extra biking and hiking and birding his retirement enables. Not that his withdrawal from Python has been carefree or complete. He likens his new relationship with the language to that of a parent dropping a kid off at college: no longer figuring in their daily life doesn't mean you stop worrying. And almost immediately after he stepped away, culture wars drew him back with a vengeance.

As we speak, Python is recovering from its lead role in a shattering pan-industry drama around the deep vocabulary of programming, specifically use of the words "master" and "slave" to describe a situation in which one process, database or device controls the action of another. Any claim to dry technical neutrality on the part of these terms became harder to sustain after a professor from Michigan traced their use in electrical engineering to 1904, the era of Jim Crow. Accordingly, an engineer named Victor Stinner, noting the racial reassessments of the Black Lives Matter movement, submitted a series of pull requests to the Python GitHub repository suggesting that "For diversity reasons, it would be nice to try to avoid 'master' and 'slave' terminology which can be associated with slavery." The git hit the fan.

Most Pythonistas agreed or at least didn't disagree with Stinner's mild observations, but a vociferous fringe—clustered on Reddit—railed against the incursion of "PCython" into their private code bubbles. A third group used more measured and engaging language to question the necessity of such a change. Temperatures rose to a level where Guido was forced to adjudicate, merging three of Stinner's four PR's and declaring the issue closed. Other languages and organizations followed suit—Rust, Go, Android, Twitter, JPMorgan Chase and most controversially GitHub, where what I've referred to as a *main* repository was traditionally called the *master* repository. Which is where the issue grew thorny. With no concomitant "slave" terminology on GitHub, *Wired* quoted one Black coder who found the move unsettling, having previously associated the term on GitHub with "master copy" or mastering a skill. Another Black coder saw the purge as vapid signaling, a distraction from the real issue of police brutality. Both had signed a petition decrying the GitHub change, although both supported dropping the word "slave" wherever it appeared in the field.

Guido imagined his work done, only to be caught off guard by a follow-up PR seeking removal of the American English style bible, *The Elements of Style* by Strunk and White, as the preferred guide to clear Pythonic communication. Seeing the style guide as incidental and understanding that nonnative English speakers may not have access to it, the Dutchman reflexively agreed to merge the request, without noticing that it wrote statements about "relics of white supremacy" into the language repo. One core developer grew so distraught and consumed with rage that he had to be asked to step down and reflect on his behavior for several months. And in a curious way I feel for that developer's distress at watching politics—*society*—begrime the pristine abstractions of his beloved language. On the other hand, a pretense that these two things are or ever could be viewed separately underlies a lot of dysfunctional code. As the Python core developer Brett Cannon notes, "I have no problem with someone wanting to keep politics out of Python, but they should recognize that this is a political position."

Why do these issues grow so heated in relation to a *computing* language? Designing a high-level "interpreted" language like Python boils down to writing rules for an interpreter that rapels code down the stack en route to becoming machine code. Do we want the interpreter to expect semicolons to delineate statements, or for indentation to do that job? Should the data type of a variable (string, integer, Boolean, etc.) be declared up front (as in C) or automatically recognized at *run time* (as in Python)? These interwoven rules become the language. Andy Hertzfeld was not being dismissive when he suggested that creating a language is more an imaginative than technical challenge. Even when he was in college, he explained, there were tools to help with the implementation of your ideas: as with any creative task, the hard part was trying to "see" something not yet there but which illuminates the world in new and useful ways.

Like Hertzfeld, Guido describes his fascination with code in terms of the way it stirs the imagination, tickling the liminal space between our corporeal experience and the cloud of abstraction summoned to represent it.

"It was these abstractions that really got my interest and kept it for forty years," he says. "I remember thinking it was so much fun."

The interesting thing is that how one represents the human environment in abstraction depends on how one sees and thinks about that environment. Once we know how languages are made and what they do, we understand why each assumes its own unique view of the world, of how things do-should-*will* work. A view of the human aspect is necessarily baked in. Even so, I've often felt a pang of guilt in admitting that I think Python reflects both the personality of its creator and a distinctively Dutch modernist aesthetic, wondering if I read too much into what is and should be a dry technical discipline.

That was before I heard about Larry Wall.

I first learned of Wall, creator of the Perl programming language, from Tim O'Reilly. My language exposure at the time was limited to Python and an enervating encounter with JavaScript. The latter seemed so anarchic and messy that I assumed it to be no more designed than a turd in the road, before learning that a man named Brendan Eich threw its rudiments together in a week. Imagine my mirth when the first item in a search on Eich concerned his well-publicized financial support for Proposition 8, a California ballot initiative seeking to re-ban same sex marriage in the state. No wonder I struggled to engage his f—ing language, I tried to resist concluding. O'Reilly smiled when I joked about the Eich discovery but said he too had long suspected programming languages—those supposed odes to detached rationality—to reflect and even embody the personalities of their makers. This was certainly true of the one he knew best, he said, who was Larry Wall.

O'Reilly majored in classics at Harvard and never set out to become the planet's premier publisher of computing books. That changed with the huge success of the now classic *Programming Perl* at a time when Perl 5, released in 1994 and soon referenced as "the duct tape of the internet," looked set to eclipse high-level rivals like Python and Ruby.

Coders like to hide the seriousness of their philosophic choices behind semi-comic acronyms. In Kent Beck's test-driven development process there is YAGNI ("You Ain't Gonna Need It") and KISS ("Keep It Simple, Stupid"), while all coders recognize TANSTAAFL ("There Ain't No Such Thing as a Free Lunch") and WYSIWYG ("What You See Is What You Get"). My own favorite motto belongs to Ruby, whose Japanese creator Yukihiro Matsumoto is known as "Matz" within the community: *MINASWAN* stands for "Matz Is Nice and So We Are Nice" and is the language's chief cultural touchstone.

Perl is more pragmatic. TMTOWTDI, spoken "Tim Toady," stands for "There's More Than One Way to Do It," a philosophy diametrically opposed to Python's founding principle that there should be only one obvious way to accomplish any task. In truth both statements represent ideals rather than hard rules, as even Guido admits to being surprised on occasion by the variety of options available to Pythonistas, while Perl's TMTOWTDI boast finally required extension to the deliciously absurd TMTOWTDIBSCINABTE ("Tim Toady Bicarbonate"), for "There's More Than One Way to Do It but Sometimes Consistency Is Not a Bad Thing Either." Which brings to mind the character in novelist Tom Drury's *Hunts in Dreams* who, spying American football for the first time, speculates that "any sport requiring so much padding had yet to arrive at an appropriate set of rules."

The point is that if Perl and Python offer a study in contrasts, their authors hail from opposite ends of any spectrum you'd care to name. Born within two years of each other in The Hague (1956) and Los Angeles (1954), respectively, they both richly deserve their

Awards for the Advancement of Free Software, but there the similarity ends. Where Guido is modest, reflective and focused, Wall is larger than life and embracing of contradiction, paradox and hyperbole. As an undergraduate he studied music and chemistry and was pre-med. He then took a few years off to explore computing, before graduating with a bespoke degree in natural and artificial languages, joining NASA when his studies were done. He is an evangelical Christian in the Church of the Nazarene but does believe in evolution. Known for his playfulness, Wall's computing talks often double as fine stand-up comedy.

From the vantage point of Guido's sofa we study one of the midcentury teak chairs around his dining table; note the way it eschews the distractions of decoration, requiring beauty and purpose to be felt simultaneously or not at all, in a way that is clear, questioning and deceptively hard to get right—as simplicity always is. He is piqued by my theory that his unconsciously modernist aesthetic, and even his Dutchness, allowed me to feel at home in a language whose assumptions were amicable to my own way of seeing (in time coding will teach me to better appreciate others, even {*get this*} JavaScript's). At Google Guido had named a pair of projects for the painter Piet Mondrian and furniture maker Gerrit Rietveld, big figures in the Dutch modernist *De Stijl* movement of the early twentieth century, while seeming never to have directly connected the substance and aesthetic of their work to his. And why would he have? Guido set out to create a language that pleased him and others. That's all. Slowly he gets what I'm saying—that his work deserves to be seen as part of a creative lineage extending far beyond code—and he starts to smile. Not that he couldn't have seen this on his own, more that he wouldn't be temperamentally inclined to entertain such a claim for himself. This is joyful to watch.

Larry Wall the Renaissance Man linguist was never innocent of his cultural bearings. In 1999, with Python lurking in the shadows and Perl on top of the world, Wall gave a talk entitled *Perl, the World's First Postmodern Programming Language*. The disquisition was long but entertaining and at times laugh-out-loud funny. It was also deadly serious, because in it Wall explicitly defines Perl against "modernist" languages like Python—which he names, claiming it to impose a way of seeing and being on the programmer, limiting freedom of expression. More broadly, he is making a point about what he sees as the constricting influence of a twentieth-century modernist aesthetic which, in venerating novelty above all else, devalues tradition and alienates the individual, is responsible for everything from ugly buildings to unlistenable music to the breakdown of the family and somewhat dull code. In questioning everything and challenging rules, he contends, modernism has simply replaced well-defined old rules with opaque new ones. Modernism is disruptive and inhumane.

Postmodernism came to prominence as a reaction to Modernism toward the end of the twentieth century. We could lose our minds trying to pin down what this intellectual movement is and incorporates (resisting definition is one of its signature tenets); but for Wall it represents a defiant eclecticism that refuses to elevate one outlook or aesthetic above another, feels free to mix cultural references at will regardless of history or context. Against Modernism's insistence on objective (if contingent) principles through which to assess value, Wall cites one of his daughter's catchphrases as capturing the spirit of his Postmodernism. "S'all good," she and her friends say, signaling *if it pleases you, it has value—we don't judge*. In this spirit he sees openness and empathy—and in naming his language for a Biblical parable ("the pearl of great value" from the Book of Matthew) he goes so far as to imply a spiritual dimension. God, he jokes, is probably a Postmodernist. Why else would He have chosen evolution as His engine of Creation?

At this point we may be wondering how any of this can possibly relate to programming. Larry is here for us: "One of the characteristics of a postmodern computer language is that it puts the focus not so much onto the problem to be solved, but rather onto the person trying to solve the problem. I claim that Perl does that, and I also claim that, on some level or other, it was the first language to do that."

He continues, "How does Perl put the focus onto the creativity of the programmer? Very simple. Perl is humble. It doesn't try to tell the programmer how to program. It lets the programmer decide what rules today, and what sucks. It doesn't have any theoretical axes to grind. And where it has theoretical axes, it doesn't grind them. Perl doesn't have any agenda at all, other than to be maximally useful to the maximal number of people. To be the duct tape of the internet, and of everything else."

What this means in practice is that Perl provides as many ways to accomplish a task as Wall and his developers have had the energy, time and ingenuity to conceive in accordance with the precept of "TimToady". Thereafter no one can tell a Perl coder how their code should look or behave. Perl's priority was and is individual freedom. And who wants to argue against freedom? JavaScript works in much the same way by default if not design.

This is compelling to me for several reasons. One, if someone had told me when I set out to program a computer that I would end up debating the merits of Postmodernism—or even being invited to type the word—I would have said they'd lost their variables. Two, my instinctual feelings about Modernism and

Postmodernism have always been precisely the opposite of Wall's. Where Wall sees Modernism as a soulless expression of arid rationalism committed to burying other perspectives, the historians who describe it as having evolved from Romanticism make more sense to me. When I look at a building by Frank Lloyd Wright or Denys Lasdun or Rosemary Stjernstedt; listen to the music of Kate Bush, John Coltrane, Autechre or David Bowie; read Hemmingway, Conrad, Forster, Lawrence or Dylan Thomas; or stand before paintings by Kandinsky, Pollock or Bacon, the feeling is of every cell in my body acquiring a kind of fluorescence, becoming intensely alive. I have no trouble understanding why others might not feel the same about these works, but for me they certainly evoke emotion. In perpetually seeking new answers to the question "*Why?*" (as in "Why are things the way they are and do they *have* to be that way?"), the Modernist standpoint will always produce a proportion of ugly or uncongenial outcomes. But to me the question always seems worth asking. And in shunning gratuitous decoration in an effort to render form and function as a unity, a Modernist artifact forces us to see what it is and is doing. Nature works the same way. Ornamentation is everywhere in the evolved world, but only where it serves function. It seems to me that this very fact, this *integrity*, comprises its fascination and beauty for us.

By contrast, I tend to experience Postmodernist products as vacuous and depressing. I could enumerate reasons for this—while also listing many, many individual exceptions (the novels of Kurt Vonnegut, the music of Moby)—but they wouldn't be important here. What does matter is that, just as Larry Wall and I might look at a building or a painting and have entirely different responses, so we do with programming languages—*for the same reasons*. This doesn't mean we can't empathize with and respect the other's perspective. We can. On the evidence I love the way Larry Wall's mind works and obviously have a lot to learn from it. At the same time, how do we explain that, more than two decades after Wall dismissed Guido's brainchild as stunted and overprescriptive while presuming his own to augur a bold postmodern future, Python is by most estimates the most popular general-purpose programming language in the world while Perl, expressive and beloved as it is to followers, bumps along the bottom of Stack Overflow surveys among the niche and legacy tongues; presents as an intriguing anachronism or cautionary tale, code's former star quarterback pumping gas at the edge of town?

Tim O'Reilly, who credits Wall as a major influence on his own thinking, generously offers to share some responsibility for Perl's fate. After the success of *Programming Perl*, he hired his friend to work full-time on the next iteration of the language, inadvertently drawing him away from the real world in the process.

"As one famous programmer said, Perl 6 is where it stopped being a working programming language and became an art project," O'Reilly tells me with a regretful shake of the head. "And it was just because Larry didn't have to solve real problems anymore."

I'm not the first person to observe that complete freedom is often the enemy of art, because constraints focus imagination. Nor that many programming languages have been written as adorable art projects, but that they tend to find little use IRL. You won't find much Whitespace, which recognizes only whitespace characters like spaces, tabs and linefeeds, at Facebook—nor Brainfuck or var'aq, a speculative language based on Klingon. Indeed, one upshot of the postmodern TimToady approach is elaborated by *The Cathedral and the Bazaar* author Eric S. Raymond, who in 2015 noted distinguished Canadian programmer Henry Spencer's likening of Perl to a "Swiss Army chainsaw," implying a language that is "highly versatile" but "ugly" and "noisy" and "distressingly inelegant." It should by now go without saying that Perlite programmers, ignoring Dijkstra's insistence that in a complex system elegance is not a luxury, "adopted [Spencer's] epithet as a badge of pride." So it is that a 2019 spoof academic paper by the ex-Google engineer Colin McMillen set out "to solve a longstanding open problem in the programming languages community," namely "Is it possible to smear paint on the wall without creating valid Perl?" The answer, according to McMillen and his coauthor "Tim Toady," was *yes*, because only 93 percent of paint spatters constitute valid Perl code. That same year Perl 6 was renamed Raku, after a form of Japanese pottery, but has gained little traction since.

None of which is to dismiss Perl. Doomed art projects often contribute important ideas to the world. Moreover, in a field boasting high levels of neurodiversity, the gift of myriad routes into and through a language can be seen as open and generous. At a conference I meet a code whiz named Evan Henshaw-Plath, known to friends as "Rabble," who was an itinerant open-source political activist before figuring this was America and he'd better earn some green if he didn't want to end up on the streets of San Francisco. Having been the first employee at the company that built Twitter, and watching in despair as social media turned toxic, he got funding to develop a new form of social network called Scuttlebutt, where the problems common to Facebook and Twitter would be designed away (and whose community happens to include an enthusiastic Nicholas). I visit Rabble at Scuttlebutt's headquarters in the Mission District and stay for a fascinating couple of hours that open with all the reasons I should dump Python for his treasured Ruby, beginning and ending with the former's supposed prescriptiveness—making religious wars the only arena in America where an anarchist and an evangelical Christian like Larry Wall can find common ground without resort to firearms. And Rabble's pitch almost works. I spend six weeks learning Ruby, which has much in common with Python, and I like it a lot, just not as much as my first language. Which is hardly surprising.

I could see an argument that Python suits a world grown more constrained and ends-related over the past few decades, while playful, freewheeling Perl does not. But I think this would be wrong: the real secret to Python's success is not the language as such, but rather the humming ecosystem of libraries written and maintained by users, some of which are deep and broad and harder to learn than the syntactic substrate. By keeping things simple and prioritizing communication, Guido allowed the creativity of a community to flourish. Against this it is tempting to see Wall's theoretically attractive choice to prioritize individual users' freedom—to "[put] the focus not so much onto the problem to be solved, but rather onto the person trying to solve the problem"—as a fascinating inquiry into a peculiarly American, abstracted notion of "freedom" as freedom *from* responsibilities to others (future readers of your code, for instance), where in most functional societies "freedom" has meant freedom *to have* those responsibilities.

Is it possible to be or aspire to be so individually free that the "freedom" we seek becomes a kind of straitjacket and means we can no longer share? Was Perl trying to warn us, back in the 1990s, that the answer could be yes? More importantly, who'd have imagined this discussion becoming crucial within the larger context of code?

Guido makes no broad claims for Python or himself, much less as against other languages. Python is his view of how programming can be done, based on his instinctual sense of how the world ought to work. In insisting that Pythonistas consider the coders who follow them, he invites them to think about those touched by their code down the line, noting that "Ideas expressed in a programming language often reach the end users of the program—people who will never read or even know about the program, but who nevertheless are affected by it."

This doesn't mean everyone will accept his invitation.

"I still don't know where it's all gonna end up," the former BDFL reflects toward the end of our conversation. "And I, I fear that five years from now I'll be utterly surprised where it went, just like I am really surprised now where things went compared to five or fifteen years ago."

He pauses, considering.

"You know, I sort of . . . I would have been satisfied with writing code for the level of technology that we had in the seventies or eighties. Right? I mean, I could see that there were always new developments and that it was moving fast, but it always stayed inside this sort of specialization of, well, there are certain things for which we need computers. And that includes lots of administration, scientific calculations and so on. That we would be using computers for social interaction to the extent that we are now? It never occurred to me."

Two revelations emerge from my encounters with acknowledged code sensei. First, even people at the epicenter of the digital revolution were blindsided by where it went. Second, of the trio I meet, all turn out to have been blessed by some combination of nature, nurture and fortune with a gift for coding that sets them apart from even highly adept peers, while also being connected, empathic, curious and fully rounded people, who exude precisely the kind of gentleness *Hacker News*'s Daniel Gackle spoke to. In other words, their sense of the rich potential of code's abstractions is reciprocated by a feel for the richness of the human environment those abstractions exist to represent. If there is an identifiable "coding mind," it is not tied to a specific personality type, much less the laconic, asocial curmudgeon of 10X mythology and PAT test presumption of Joseph Weizenbaum's "obsessive coder" nightmares. In fact, the code savant stereotype—of someone who can parse syntax and process coded abstraction with the ease of sifting sand but to whom the human realm is also processed as a kind of abstraction—may provide one high-level definition of a bad, potentially dangerous coder.

So why does the myth persist?

This is not a rhetorical question.

CHAPTER

13

Enter the Frankenalgorithm

This is a special way of being afraid
No trick dispels. Religion used to try,
That vast moth-eaten musical brocade
Created to pretend we never die,
And specious stuff that says No rational being
Can fear a thing it will not feel, *not seeing*
That this is what we fear—no sight, no sound,
No touch or taste or smell, nothing to think with,
Nothing to love or link with,
The anaesthetic from which none come round.

—Philip Larkin, *Aubade*

The 18th of March, 2018, became a date tech insiders had been dreading. That night a new moon added no light to a poorly illuminated four-lane road in Tempe, Arizona, as a specially adapted Uber Volvo XC90 detected an object ahead. Part of a digital gold rush to develop self-driving vehicles, the SUV had been proceeding autonomously for almost twenty minutes with no input from its human backup driver. Now an array of radar and light-emitting "lidar" sensors allowed onboard algorithms to calculate that, given their host vehicle's steady speed of 43 miles per hour, the object was six seconds away—assuming it remained stationary. But objects on roads seldom remain stationary, so more algorithms crawled a database of recognizable mechanical and biological entities, searching for a fit with this one's behavior.

At first the computer drew a blank. Recalculation seconds later suggested another car, predicted to drive away and require no special action. Only at the last second was a clear identification found—a woman with a bike, shopping bags dangled from handlebars, doubtless assuming the Volvo would route around her as any ordinary vehicle would. But this one didn't. Barred from taking evasive action on its own, the computer abruptly handed control back to a human master who was gazing not up but down, lost in the infinity of her smartphone. Elaine Herzberg, aged forty-nine, was struck and killed.

Worse was to come.

Seven months after the Uber tragedy, on October 28, a Lion Air Boeing 737 MAX 8 jetliner took off from Bali in Indonesia, headed for Jakarta. The new plane was barely clear of the runway when its control column began to convulse, warning that a stall could be imminent despite no other evidence of danger. Three minutes in and with the pilots still puzzling, the onboard computer seized control and drove the plane's nose downward, plunging the beleaguered vessel 700 feet toward the ground. The pilot wrestled the nose back up, only for the computer to reengage and force another steep dive ten seconds later, a man-machine battle that raged for another nine minutes with only one likely winner as the desperate crew strained and pulled until, exhausted, they could pull no more and entered a final dive from 5,000 feet, striking the Java Sea at 450 miles per hour and disintegrating, killing everyone onboard.

In the years running up to the Lion Air disaster, air travel had established itself as a miracle of safe transit. Crashes were rare. This loss was shocking. Which is why Boeing first sought to blame the crew, sowing a narrative of ethnic incompetence as cause. But when an Ethiopian Airlines flight out of Addis Ababa went down in similar circumstances five months later, media reporting and regulatory investigation began to focus on a more dire possibility. Could these 346 deaths have been caused by the 737's *software*? If so, more reflective members of the tech community would be left with a further pair of uncomfortable questions. Had these algorithmic accidents been inevitable? And were they, in the traditional sense, *accidents*?

The Uber and Boeing calamities look qualitatively similar. Flawed software caused transport vehicles to behave erratically and kill people. Dig deeper, however, and they diverge as night from day. When detailed reporting of the 737 MAX 8's fall from grace appeared, backed by the Federal Aviation Administration's own two-year investigation, the narrative arc was clear. Spooked by an advanced new offering from its European competitor Airbus, Boeing used commercial muscle to buy a shortcut. Rather than go to the trouble and expense of designing a new plane and having to retrain pilots, engineers affixed new engines to an old one. But these were larger engines for which that plane had not been intended; that rendered it vulnerable to stalling. And just as musicians joke about taking a subpar studio performance and "fixing it in the mix," so Boeing sought to patch slack design with code.

There were supposed to be guardrails. The Federal Aviation Authority previously stood as a beacon of careful software management, insisting that every line of code in a safety-critical program be traceable to a specific requirement in the design document, allowing no space for unplanned algorithmic entanglement. But the FAA had grown cozy with Boeing, a company historically run by engineers but now in the hands of bean counters teabagging the bottom line, prioritizing shareholder value over safety. Where pilots should have been trained in the 737's revised flight control system, the FAA granted permission to bypass this expensive process and even decline to inform pilots of the change. A 2021 Justice Department criminal settlement established that Boeing employees knowingly withheld information about the new software from regulators. A faulty sensor in each lost plane was enough to seal its fate. New software had been asked to provide a fix-in-the-mix for bad engineering and lax thought. It couldn't. And never can.

Few outside the software industry will have realized, but the lonely Uber death was more portentous than Boeing's splashier failing. At root we know what an algorithm is by now: a small, simple thing; a rule for treating data. If *a* happens, then do *b*; if not, do *c*. If the user claims to be eighteen, let them into the site; if not, print "Sorry, you must be 18 to enter." Al-Khwarizmi would recognize this definition. If/then/else. Easy. Computer programs are often described as bundles of such algorithms, recipes for processing input.

We no longer accept the sales pitch for this classic form of algorithm so meekly as we did. In her 2016 book *Weapons of Math Destruction: How Big Data Increases Inequality and Threatens Democracy*, Cathy O'Neil, a math prodigy who left Wall Street to teach and run her mathbabe blog, demonstrated beyond doubt that, far from eradicating human biases as tech companies liked to claim, algorithms could magnify and entrench them. As we know, software is overwhelmingly written by affluent white and Asian men and will unavoidably reflect their assumptions. A slew of books and academic papers expanded on O'Neil's thesis, among them Virginia Eubank's *Automating Inequality: How High-Tech Tools Profile, Police, and Punish the Poor* and Safiya Umoja Noble's *Algorithms of Oppression: How Search Engines Reinforce Racism*. But the effects can also be subtle. So it is that we've found outstanding teachers fired by algorithmic ranking and popular TV shows thought to have been canceled in the same way; students learning to game algorithms that allocate grades or being algorithmically marked down in important exams for having the temerity not to attend a fee-paying school. We've seen manufacturing production lines break or cause injury through *over*-automation, citizens misidentified as criminals or terrorists and arrested or refused visas, Black people served ads assuming arrest records where none exist.

These algorithms were and are hubristic, ill-considered or poorly executed, part of a broad push to increase "productivity"—redefined as corporate stock price—with no expenditure of effort, imagination or capital, simply by removing employees from the equation. Hundreds of examples of this type of blunt decision-making algorithm are now monitored on the AI Incident Database and elsewhere. "Often I'll speak with my fellow engineers and they'll have an idea that is quite smart," the database's founder, a software engineer named Sean McGregor, told *Wired*. "But you need to say, 'Have you thought about how you're making a dystopia?'" By August 2021 *MIT Technology Review* would be producing a podcast on how to "beat" job recruitment algorithms their researchers had identified as reductive, inhumane and far harder to correct than might be imagined.[1] Most of the problematic algorithms in this category could best be improved by not being written. Yet they *are* being written, flung like digital spanners into the works of everything from healthcare to law, human rights,

1 Hilke Schellmann, "Auditors Are Testing Hiring Algorithms for Bias, but There's No Easy Fix," *MIT Technology Review*, February 11, 2021, https://www.technologyreview.com/2021/02/11/1017955/auditors-testing-ai-hiring-algorithms-bias-big-questions-remain/.

and real estate (where an app to aid property flipping is unlikely to add to the sum of human happiness).

We might call these workaday algorithms "fixed" or "dumb" in the sense that they're working to known formulas, treating data input according to "rules" defined by humans. Changes to these programs involve programmers rewriting code and their quality of output depends on the care with which this is done. There are grounds for optimism here. While we don't always know where these algorithms are, we do at least know *what* they are. Cathy O'Neil has called for "algorithmic audits" of decision-influencing systems that affect the public—and the good news is this battle is underway. The bad news is that it already looks quaint in relation to where we're headed.

Since the rise of the internet (and search engines in particular), popular usage of the term "algorithm" has changed in meaningful ways. Right now my iPhone plays a podcast in which journalists dissect the latest scandal engulfing Facebook and its parent company Meta, in a way that's become a near constant backdrop to twenty-first century life. Yet my ear is drawn less to details of the case than how participants return over and again to "the Facebook algorithm" as a monolithic entity, in a way that recalls Mary Shelley's writing about the monster in *Frankenstein*, as if it were sentient and had a life of its own.

Until recently this could have been dismissed as mere algopomorphic misperception, ascribing human characteristics to an algorithm. No longer. In theory, modern algorithms exist on a spectrum running from the classic fixed code programs discussed so far to the more or less distant dream of artificial human-like intelligence, sometimes referred to as artificial general intelligence

(AGI). The problem is this: between the "dumb" fixed algorithms and holy grail of true AGI lies a problematic halfway house we entered a decade ago with almost no debate, much less agreement as to aims, ethics, safety or best practices. Recent years have seen these new algorithms marketed under the blanket term "AI," but this was not always the case.

What we now call "AI" is arguably better described as machine learning. The many competing branches and techniques go in and out of fashion but have one thing in common: they are founded on algorithms empowered to "learn" from their environment by writing their own rules of engagement with it. Which means they write their own code. An autonomous driving algorithm, shown enough stop signs, will amass a sophisticated set of rules for recognizing them. A "language model," fed enough text, learns which words are most likely to follow each other in any given context, and the more text it encounters, the more subtle its sensitivity to the totality of each word's context, so the more plausibly it simulates human intelligence, despite being little more than "autocomplete on steroids" in the words of one skeptical expert. In each case the algorithms are working statistically—making, reinforcing and weighing connections between things on the basis of probability and pattern recognition rather than understanding. By the time language models alert the general public to the power and ambiguities of "AI"-branded machine learning algorithms at the end of 2022, they are neither new nor unique. Rather, they are the tip of a very large iceberg.

Can these machines be accurately described as intelligent? A properly intelligent machine would be able to question the quality of its calculations based on something like our own intuition, which we might think of as a broad accumulation of knowledge and experience drawn from living and moving through the world. Pivotal to our intuitive sense is an ability to transfer lessons from one context to another, so-called *transfer learning*, the form of "general" intelligence that enables us to say "this calculation doesn't look right in light of everything I know, therefore I'll revisit it." At this time of writing no machine comes close to demonstrating this kind of intelligence, which makes them less generally capable than a toddler, a crow, a cuttlefish or a bee. The difficulty is that once an algorithm is learning, which means *writing its own rules*, we no longer know to any degree of certainty what those rules are . . . at which point we can't be sure how it will interact with other algorithms, or the physical world, or us. Is it overdramatic to call these code chimeras "Frankenalgorithms"? Maybe. Maybe not. That's just the thing: at present we can't know. We're placing bets. Letting the wheel spin. Welcome to the exciting, unnerving world of "AI."

How should we think about "AI"? I first encountered these algorithms in 2013 while researching high-frequency trading (HFT) on the stock market. What I found was earthshaking: a human-made digital ecosystem, distributed among racks of black boxes crouched like ninjas in billion-dollar data farms, which is what stock markets had become—each with an identical multibillion-dollar backup facility waiting to take over in the event of outage. Where once there had been a physical trading floor, action had devolved to a central server through which nimble, predatory algorithms fed off lumbering institutional ones, tempting them to sell low and buy high by fooling them as to the state of the market. Although no human actively traded anymore, HFT masters called the large, slow participants "whales," and they mostly belonged to mutual and pension funds, i.e. the public. For most HFT shops, whales were now the main profit source. In essence these algorithms were trying to outwit each other; were doing invisible battle at dizzying speed, placing and canceling the same order

10,000 times per second or slamming so many into the system that the whole market shook—all beyond the oversight or control of people. The real-world value of a company figured nowhere in this new market matrix.

No one could be surprised that this situation was unstable. A "flash crash" had occurred in 2010, during which the market went into freefall for five traumatic minutes then righted itself over five more—for no apparent reason. I traveled to Chicago to see a man named Eric Hunsader, whose prodigious programming skills allowed him to see market data in far more detail than regulators, and he showed me that by 2014, "mini flash crashes" were occurring every week. Even he couldn't prove exactly why these shudders happened, but he and his staff had begun to name some of the "algos" they saw, much as crop circle hunters named the formations found in English summer fields, dubbing them "Wild Thing," "Zuma," "The Click," "Disruptor."

Professor Neil Johnson, a physicist specializing in complexity at George Washington University, coauthored a study of stock market volatility. "It's fascinating," he told me. "I mean, people have talked about the ecology of computer systems for years in a vague sense, in terms of worm viruses and so on. But here's a real working system we can study. The bigger issue is that we don't know how it's working or what it could give rise to. And the attitude seems to be 'out of sight, out of mind.'"

Significantly, Johnson's paper on the subject was published in the journal *Nature* in 2013 and described the stock market in terms of "an abrupt system-wide transition from a mixed human-machine phase to a new all-machine phase characterized by frequent black swan events with ultrafast durations." I spoke to the science historian and computing theoretician George Dyson (last seen discussing ENIAC in chapter seven), who told me the scenario was complicated by the fact that some HFT firms were allowing the algos to learn—"just letting the black box try different things, with small amounts of money, and if it works, reinforce those rules. We know that's been done. Then you actually have rules where nobody knows what the rules are: the algorithms create their own rules—you let them evolve the same way nature evolves organisms." Non-finance industry observers (and a few insiders) began to postulate a catastrophic global "splash crash," while the fastest-growing area of the market became instruments that profit from volatility. In his 2011 novel *The Fear Index*, Robert Harris imagines the emergence of AGI—of the Singularity, no less—from precisely this digital ooze. To my surprise, no scientist I spoke to would rule such a possibility out.

All of which could be dismissed as high finance arcana were it not for one simple fact. Wisdom once held that all new technology was adopted first by the porn industry, then by everyone else. But the twenty-first century's porn is finance, so when I thought I saw signs of HFT-like algorithms causing problems elsewhere a few years later, I called Neil Johnson again.

"You're right on point," he told me. A new form of algorithm was moving into the world, which in being allowed to evolve under its own steam was analogous to "a genetic algorithm." He thought he saw evidence of them on fact-finding forays into Facebook ("I've had my accounts blocked four times," he chuckled). If so, algorithms are jousting there, and adapting, as on the stock market.

"After all, Facebook is just one big algorithm," Johnson said. "And I think that's exactly the issue Facebook has. They can have simple algorithms to recognize my face in a photo on someone else's page, take the data from my profile and link us together. That's a very simple concrete algorithm. But the question is what is the effect of billions of such algorithms working together at the macro level? You can't predict the learned behavior at the level of the population from microscopic rules. So, Facebook would claim that they know exactly what's going on at the micro level, and they'd probably be right. But what happens at the level of the population? That's the issue."

To underscore this point, Johnson and a team of colleagues from the University of Miami and Notre Dame produced a paper, "Emergence of Extreme Subpopulations from Common Information and Likely Enhancement from Future Bonding Algorithms" (2018), which, in addition to providing an airtight test for sobriety after a night on the town, purports to mathematically prove that efforts to "connect" people on social media will always polarize a population. Johnson has been studying this issue from an interdisciplinary perspective for years, but by July 2021 a diverse international group of scientists and social scientists, writing in the august *Proceedings of the National Academy of Sciences of the United States of America*, would be considering the social effects of technology in explicitly biologic and human evolutionary terms, positing its study as a "crisis discipline" like climate science.

Cathy O'Neil tells me she consciously excluded this adaptive form of algorithm from *Weapons of Math Destruction*, because apportioning responsibility to a particular phalanx of code becomes difficult to impossible in a convoluted algorithmic environment where nothing is clear. This opacity makes them easier to miss, dismiss or ignore, she adds, before advising that if I want to see them in action I should ask what a flash crash on Amazon might look like.

"I've been looking out for these algorithms, too," she says, "and I'd been thinking: 'Oh, Big Data hasn't gotten there yet.' But more recently a friend who's a bookseller on Amazon has been telling me how crazy the pricing situation there has become for people like him. Every so often you will see somebody tweet 'Hey, you can buy a luxury yarn on Amazon for $40,000.' And whenever I hear that kind of thing, I think: 'Ah! That must be the equivalent of a flash crash!'"

Anecdotal evidence of anomalous events on Amazon is plentiful in the form of threads from bemused sellers and at least one academic paper from 2016. The latter claims: "Examples have emerged of cases where competing pieces of algorithmic pricing software interacted in unexpected ways and produced unpredictable prices, as well as cases where algorithms were intentionally designed to implement price fixing." The problem, again, is how to serve responsibility in a chaotic habitat where simple cause and effect either doesn't apply or is nearly impossible to track. Frances Haugen's 2021 whistleblowing disclosure that Facebook long knew their algorithms caused harm could imply one of two things: one, the corporate central committee didn't care, or two, it cared but didn't know what to do short of shutting down the algorithm—that is, shutting down *themselves*. The second possibility is the more alarming.

George Dyson anticipated much of what is happening today in his classic 1997 book *Darwin among the Machines*. The problem, he tells me, is that we're building systems that are beyond our intellectual means to control. We believe that if a system is deterministic (acting according to fixed rules, this being the definition of an algorithm) it is predictable, and that what is predictable can be controlled. Both assumptions turn out to be wrong.

"It's proceeding on its own, in little bits and pieces," he tells me. "What I was obsessed with twenty years ago that has completely taken over the world today are multicellular, metazoan digital organisms, the same way we see in biology, where you have all these pieces of code running on people's iPhones, and collectively it acts like one multicellular organism."

He continues, "There's this old law called Ashby's law that says a control system has to be as complex as the system it's controlling, and we're running into that at full speed now, with this huge push to build self-driving cars where the software has to have a complete model of everything, and almost by definition we're not going to understand it. Because any model that we understand is gonna do the thing like run into a fire truck 'cause we forgot to put in the fire truck."

Or the bike with shopping bags draped over handlebars. Unlike our old electromechanical systems, these new algorithms, whatever we choose to call them, are also impossible to test exhaustively: unless and until we have superintelligent machines to do this for us, we're going to be walking a tightrope. On September 21, 2021, the *New York Times* ran an article entitled "The Costly Pursuit of Self-Driving Cars Continues on. And on. And on." The piece was subtitled "Many in Silicon Valley promised that self-driving cars would be a common sight by 2021. Now the industry is resetting expectations and settling in for years of more work." In a pattern endemic to the venture-capitalist-driven tech market in general, most of the early self-driving car boosters have now dropped out, understanding how much more difficult the task is than they imagined—a fact that should ring alarm bells in itself. Even Tesla caused a stir in April 2021 by allowing for the first time that it may not produce self-driving cars "in the timeframe we anticipate, or at all," moving into hot water with regulators by the end of the year. Uber bailed in 2020, by some accounts running a poetically pleasing coach and horses through their business model in the process. When Amazon bought the San Francisco "autonomous taxi" startup Zoox in 2020 as the young firm was running out of cash, it was hard not to notice that while its freshly launched vehicle looked innovative, the big question—how the hell its code would work—was glossed over.

Dyson first told me in 2018, at the peak of self-drive mania, that he doubted we would ever see autonomous cars roam freely through city streets. Now Toby Walsh, a distinguished professor of artificial intelligence at the University of New South Wales in Sydney, Australia (and one of those code savants who wrote his first program at thirteen and ran a fledgling computer business by his late teens), explains from a technical perspective why that might be.

"No one knows how to write a piece of code to recognize a stop sign. We spent years trying to do that kind of thing in AI—and failed! It was rather stalled by our stupidity, because we weren't smart enough to learn how to break the problem down. You discover when you program that you have to learn how to break the problem down into simple enough parts that each can correspond to a computer instruction [to the machine]. We just don't know how to do that for a very complex problem like identifying a stop sign or translating a sentence from English to Russian—it's beyond our capability. All we know is how to write a more general-purpose algorithm that can learn how to do that given enough examples."

Hence the emphasis on machine learning techniques (loosely) based on the brain's neural networks. But until our machines are intelligent enough to understand *why* they do what they do—in other words, are *actually* intelligent—we will be empowering algorithmic systems that write themselves uncritically, and are understood by nothing and no one. If these systems are undertaking mechanical or data-sifting tasks and seem to work, like searching for patterns in medical data or allowing us to see a damaged Rembrandt as the master intended, we may decide to embrace them. If we are cavalier enough to allow this type of evolving algorithm to intervene in something as complex as human relations, at scale, then we may have to accept that if our machines are not yet intelligent, neither are we. And yet here we are. Ellen Ullman elucidated the problem when I spoke to her as long ago as 2017, saying:

"When programs pass into code and code passes into algorithms and then algorithms start to create new algorithms, it gets [further and further] from human agency. People say, 'Well, what about Facebook—they create and use algorithms and they can change them.' But that's not how it works. They set the algorithms off and they learn and change and run themselves. Facebook intervenes in their running periodically, but they really don't control them. And particular programs don't just run on their own, they call on libraries, deep operating systems and so on."

In the coauthored words of the Cornell University network and algorithm specialist Jon Kleinberg, "We have, perhaps for the first time ever, built machines we do not understand." For Alan Turing this was always the goal: a lecture he gave in London in 1947 included the first known public mention of computer intelligence, offering that "What we want is a machine that can learn from experience" through being allowed to "alter its own instructions." Indeed, the great logician's Bletchley Park and Manchester University colleague Max Newman claimed his friend Turing was talking and thinking about AI from the very beginning, in the 1930s. Now the endgame has begun—and is likely to accelerate, according to Toby Walsh:

"We will eventually give up writing algorithms altogether, because the machines will be able to do it far better than we ever could. Software engineering is in that sense perhaps a dying profession. It's going to be taken over by machines that will be far, far better at doing it than we are."

Walsh believes this makes it more, not less, important that the public learn about code, because the more alienated we become from it the more it seems like magic beyond our ability to affect. When shown the definition of "algorithm" I provided earlier, he finds it incomplete, reprising Ellen Ullman in saying: "I would suggest the problem is that 'algorithm' now means any large, complex decision-making software system and the larger environment in which it is embedded, which makes them even more unpredictable." A provocative thought, which leads him to foresee ethics as the next frontier in tech, leading to "a new golden age for philosophy." Professor Eugene Spafford of Purdue University, a cybersecurity expert, spells out for me one of the knotty questions ethicists will face.

"Where there are choices to be made, that's where ethics comes in. And we tend to want to have an agency that we can interrogate or blame, which is very difficult to do with an algorithm. This is one of the criticisms of these systems so far, in that it's not possible to go back and analyze exactly why some decisions are made, because the internal number of choices is so large that how we got to that point may not be something we can ever recreate to prove culpability beyond doubt."

The counterargument is that, once a program has slipped up, the entire population of programs can be rewritten or updated so it doesn't happen again—unlike humans, whose propensity to repeat mistakes will doubtless fascinate intelligent machines of the future. Against the benefits of this flexibility we should note that, while reconfiguring the operating system of a car or medical device may be relatively straightforward, the same is not always true of people, so the flexibility argument is unlikely to hold for machines engaged in anything social. Furthermore, while automation of mechanical tasks may be safer in the long run, our existing system of tort law, which requires proof of intention or negligence, will need to be rethought. A dog is not held legally responsible for biting you; its owner might be, but only if the dog's action is thought to have been foreseeable. In our new algorithmic environments, many unexpected outcomes may not have been forecastable by people—a feature with the potential to form a scoundrel's charter whereby deliberate obfuscation becomes at once easier and more rewarding. Culpable corporations have benefited from the cover of complexity for decades (see the cases of thalidomide or Monsanto's Roundup weedkiller), but with evolved code the consequences could be harder to reverse.

And yet even the above examples may look like mere specks in the algozoan ooze before long.

🐞 🐞 🐞

If the military no longer drives innovation to the degree it once did, it remains tech's most consequential adopter. No surprise, then, that an outpouring of concern among scientists and some tech workers accompanied revelations that autonomous weapons are ghosting toward the battlefield in what amounts to an algorithmic arms race. Since the turn of the century, robotic Samsung SGR-A1 sharpshooters have policed the demilitarized zone between North and South Korea, and while their manufacturer denies that they are capable of autonomy, this claim is widely disbelieved. Russia, China and the United States all profess to be at various stages of developing swarms of coordinated, weaponized autonomous drones, while the latter may already have missiles capable of hovering over a battlefield for days, observing and *learning* before selecting their own targets. In May 2021 the United Nations Security Council reported what may have been the first instance of AI-driven lethal autonomous military drones hunting down humans, causing "significant casualties" in the Libyan Civil War. Some experts claimed this was anything but new.

And where the military goes, domestic law enforcement tends to follow: in December 2020 the *New York Times* ran a piece under the shocking headline "Police Drones Are Starting to Think for Themselves."

Algorithmic weapon programs have not gone unchallenged. As previously noted, in 2018 a group of Google employees resigned and many more protested over the tech monolith's provision of machine learning software to the Pentagon's Project Maven "algorithmic warfare" program. Management eventually responded by agreeing not to renew its Maven contract, and under duress publishing a code of ethics for the use of its algorithms. At this writing competitors including Amazon and Microsoft have shown no inclination to follow suit.

Like other "AI" agents, Google had claimed moral virtue for its Maven software; that it would help choose targets more efficiently and thereby save lives. The killing of ten members of an innocent Afghan family—including seven children—by a US drone meant for ISIS insurgents at the end of 2021 brought fresh revelations about the dangerous inefficiency of these weapons, a context in which Big Tech's offer to fix it in the mix with "AI" must look tempting.

The problem for coders of "AI" sold to the military is this: How can they or their managers presume to know what their algos will do in the field? Given the certainty that all sides will develop adaptive algorithmic countersystems designed to confuse enemy weapons, unpredictability is likely to be seen as an asset rather than handicap, as with HFT in the stock market. In this and other ways we risk in effect turning our machines inside out, wrapping the world we move through day-to-day in spaghetti code, the way one Python or Perl function might be "wrapped" within another. This is code we neither fully comprehend nor direct.

Lucy Suchman, professor of anthropology of science and technology at Lancaster University in the United Kingdom, has testified before the United Nations on autonomous weapons. She also coauthored an open letter from technology researchers to

Google, asking them to reflect on the rush to militarize their work. Tech firms' motives are easy to fathom, she says: military contracts have always been lucrative. For the Pentagon's part, she suggests that a vast network of sensors and surveillance systems has run ahead of any ability to organize and use the screeds of data so acquired.

"They are overwhelmed by data, because they have new means to collect and store it, but they can't process it. So, it's basically useless—unless something magical happens. And I think their recruitment of Big Data companies is a form of magical thinking in the sense of 'here is some magic technology that will make sense of all this.'"

Algorithmic weapons promise not just to put these oceans of data to use, but to automate decision-making based on it.

"At which point it becomes even less accountable and open to questioning. It's a really bad idea."

Suchman's colleague Lilly Irani, a former computer scientist and Google employee turned professor of communication and science studies at the University of California San Diego, reminds us that information travels around an algorithmic system at close to the speed of light, free of human oversight. Technical discussions are often used as smokescreens to avoid responsibility, she suggests.

"When we talk about algorithms, sometimes what we're talking about is bureaucracy. The choices algorithm designers and policy experts make are presented as objective, where in the past someone would have had to take responsibility for them. Tech companies say they're only improving accuracy (i.e., the right people will be killed rather than the wrong ones) and in saying that, the political assumption that those people on the other side of the world are more killable, and that the US military gets to define what suspicion looks like, goes unchallenged. So technology questions are being used to close off some things that are actually political questions. The choice to use algorithms to automate certain kinds of decisions is political too."

The problem Irani describes turns out to be a general feature of the algosphere. Cathy O'Neill and others point out that decision-automating software related to things like loans, benefits, education and crime prevention disproportionately affects low-income groups, because affluence still buys the attentions of *people*, who are compelled to own and explain judgments. In the life-and-death realm of war, legal convention assumes human accountability for decisions made. At the very least, algorithmic warfare muddies the water in ways we may grow to regret.

Solutions exist or can be found for most of the problems described here, but not without incentivizing Big Tech to place the health of society above its bottom lines. A snowballing list of suggestions as to how this might be accomplished must start with recognizing online social activities as safety-critical in the way of transport, medicine and the environment, and acknowledging that any meaningful distinction between our "virtual" and physical lives has collapsed. More serious in the long term is growing conjecture that current classical programming paradigms are no longer fit for purpose given the size and interdependency of the algorithmic systems we depend on. One solution is the rigid mapping of every coded instruction to a specific design spec, as historically practiced by the Federal Aviation Administration, but this is impractical at scale. Alternatively, some claim solutions might be found in an ultra high-level technique employed in aerospace, called model-based programming, in which machines do most of the coding and can test as they go.

Model-based programming may not be the panacea some hope for, however. Not only does it push humans yet further from the process by expanding the stack of abstraction, but the aforementioned physicist Neil Johnson conducted a study for the Department of Defense that found "extreme behaviors that couldn't be deduced from the code itself" even in systems built using this technique. Much energy is being directed at finding ways to trace unexpected algorithmic behavior back to the specific lines of code that caused it. No one knows if a solution (or solutions) will be found, but none are likely to work where aggressive algos are *designed* to clash with and/or adapt to each other.

As we wait for a technological or regulatory answer to the problem of soaring algorithmic entanglement, what are the precautions we can take? Paul Wilmott, a British expert in quantitative analysis (a "quant") who was closely involved in the battle against financial meltdown in 2008 and is a vocal critic of high-frequency trading on the stock market, wryly suggests "learning to shoot, make jam, and knit." More practically, the software security expert Eugene Spafford advises making tech companies responsible for the actions of their products, whether specific lines of rogue code can be identified or not. He notes that the venerable Association for Computing Machinery has updated its code of ethics along the lines of medicine's Hippocratic oath, mandating that computer professionals do no harm and consider the broader impacts of their work. Neil Johnson, for his part, considers our algorithmic discomfort to be at least partly conceptual, growing pains in a new realm of human experience. He laughs in noting that when he and I first spoke about this stuff a few short years ago, my questions were niche concerns, restricted to a few people who pored over the stock market in unseemly detail.

"And now, here we are—it's even affecting elections. I mean, what the heck is going on? I think the deep scientific thing is that software engineers are trained to write programs to do things that optimize—and with good reason, because you're often optimizing in relation to things like the weight distribution in a plane, or a most fuel-efficient speed: in the usual, anticipated circumstances optimizing makes sense. But in unusual circumstances it doesn't, and we need to ask: 'What's the worst thing that could happen in this algorithm once it starts interacting with others?' The problem is we don't even have a word for this concept, much less a science to study it."

He pauses a moment, trying to wrap his brain around the problem.

"The thing is, optimizing is all about either maximizing or minimizing something, which in computer terms are the same. So, what is the opposite of an optimization (i.e., the least optimal case), and how do we identify and measure it? The question we need to ask, which we never do, is: 'What's the most extreme possible behavior in a system I thought I was optimizing?'"

Another brief silence ends with a hint of surprise in Johnson's voice.

"Basically, we need a new science," he says.

I spend a long time thinking about what this science of worst possible outcomes might be called.

CHAPTER

14

ALGORAVE?

Music is a hidden arithmetic exercise of the soul, which does not know that it is counting.

—Gottfried Leibniz

The breakthrough came, as they often do, when I had all but given up. From the dawn of my interest in code, experienced advisors had asked variants of the same question. What do you want to do? What do you want to make/play/see on your screen? But my path to code had been unusual, based on an inversion of the norm. I didn't want to learn code so I could make things, I wanted to make things as a way of understanding how code was reshaping my world, was redrawing the relationships through which I understand myself to live. Yet my code confidants knew progress without practical goals would be hard, like a child trying to learn soccer skills on their own. You can go only so far before you need to join a game.

For over a year I had been focused on the low-level weirdnesses of the machine while picking through high-level basics in a way that felt dutiful but lacking a key imponderable. *Passion.* And in the end, I let go. If I found something to excite me with the locomotive force I saw in so many of the coders I now knew and admired, I would embrace it. I would also try to make peace with the possibility this might not happen, that I might be made differently from Sagar and Nicholas and all the others, remaining an appreciative tourist in their milieu, the kid in the candystore who can't reach the candy. Until, without warning, my coding ennui was shattered by a pair of discoveries that turned my outlook upside down.

I'd encountered the idea of "civic code" through scattered voluntary groups and projects. But Code for America was something different, a well-run, well-funded nationwide organization, sometimes likened to a digital Peace Corps, with "brigades" of volunteers in most major US cities. I watched a gripping TED talk by CfA's founder, Jennifer Pahlka, then went to see her at the organization's HQ on Mission Street in San Francisco.

Pahlka is a remarkable person. From a start in the game industry, she rose to be America's deputy chief technology officer under Barack Obama, founding the United States Digital Service to bring consistency to government software activities. As we move through a bright and airy open-plan office in search of a side-room that *isn't* occupied by groups of mostly young engineers in animated conversation, she explains the insight that drove her to found both the US Digital Service and Code for America, starting with the last universally acknowledged truth in America: historically, government software has tended to suck. Clunky code can be frustrating in itself, Pahlka notes, but the real tragedy comes as a secondary effect. While botched development can cost money, bad user experience costs *trust*. At a time when faith in democracy and government is under siege, every disagreeable interaction with officialdom becomes magnified and increases mistrust. To build the kind of government a flourishing democracy needs, in which the best coders are proud to serve, those bad interactions needed to turn good.

Pahlka saw that the increment between bad and good was often small. That the benefits of well-designed software could be dramatic. Simple digital tools could lessen the burden on government by empowering citizens to act for themselves; could organize crazed data and make visible the obscure, encouraging transparency and helping dismantle bureaucratic roadblocks to the things people need, creating better government in the process. The trick, she suggests, is to think of government as a platform to deliver those things we can't do or get for ourselves. It should act less like a private company (least of all a tech company) than the internet itself—"permissionless, open, generative." Government is not the same as politics, Pahlka pointed out as we sipped coffee, occasionally interrupted by CfA staff needing her attention. At root government is bureaucracy: the art of getting stuff done. Pahlka's job is to make bureaucracy sexy to a generation raised on the internet and accustomed to proactivity. And I think she may be the only person on Earth who can make this sound both appealing *and* possible.

I joined the Oakland brigade, OpenOakland, which met every Tuesday in a wood-paneled room at the magnificent Beaux-Arts-style City Hall, and over the space of a year reveled in watching exceptional things being made by ad hoc teams of developers at all points on the experience spectrum. As with all events involving coders, proceedings began with takeout food appearing as if by magic. Skewed male, white and Asian at first, the group diversified over the months, a consequence of outreach and—I couldn't help suspecting—the first slow stirrings of a shift within the industry. Favorite meetings were always the monthly Project Nights, where teams described their progress and called for any extra skills they might need, which could involve languages like Ruby or JavaScript, but also design, research, content editing or marketing. Fresh projects were often pitched on these nights, sometimes by new or existing members of the group, other times by local nonprofits or city departments. If enough attendees were interested in forming a team around the proposal, it flew.

The OpenOakland atmosphere was always light and fun and over the months I grew fond of the group. Not yet confident enough to offer code, I watched and waited and tried to be useful where possible, getting a first sense of how the elements of a software project fit together, of the daunting responsibility and even *courage* involved in sending something into the world that might appear small yet have enormous impact on whoever it touched. There were projects on making local election campaign finance data easier for citizens to access and interpret; on logging Oakland's renowned collection of murals and street art as it came under threat from gentrification; on helping California's most multinational county, Alameda, prepare for a first digital census in a community where over one hundred languages are spoken; on assisting a nonprofit to interpret reams of data on California's uneven drinking water quality, previously buried in spreadsheets and all but useless, by providing a suite of stunning graphical tools that allowed researchers and citizens alike to see patterns, trends and dangers for the first time . . . a simply incredible feat. Veering between now familiar thoughts that I was an impostor who shouldn't be there and wishful optimism that I would eventually be of use, I found myself yearning for skills in a way reminiscent of learning musical instruments as a child, when the fantasy of being able to play a particular piece could feel so intense it hurt. Little did I know these two longings were about to collide.

In retrospect it's odd that I never imagined there were musical subcultures within programming, odder still that I learned of their existence by chance. Like a lot of families, music is a shared pleasure in mine. Computer programming is not. So the last source I expected for a revolution in my code life was my poet daughter, Lotte. "Dad, there's an event happening in London that you might want to look into." It was an evening of "algorave" she said.

Algorave? I went online and could hardly contain my excitement. Here was something I never thought to see: a vibrant underground coding community—a nascent musical *scene*, no less—gathered around a new artform, of which algorave was an offshoot. This new form was called "live coding" and involved the performance of music in real time through direct manipulation of code. Wheeling around TOPLAP, an electronic zine-cum-clearing house based in Sheffield in the United Kingdom, live coding proved to be cultish and yet impressively global. I contacted TOPLAP's organizer, a digital musician, artist and researcher named Alex McLean, who told me about a showcase coinciding with my next visit to London, at the endearingly begrimed Corsica Studios in Elephant and Castle. We arranged to meet the next morning. I couldn't wait. *Finally*, I thought, *I've found a project: the thing I want to do with code.*

Most people agree that we're living through a technological revolution. We can be forgiven for thinking this is about the internet. But it's not. When the first edition of *Wired* magazine appeared in 1993, the net was mentioned just once, in passing. The revolution *Wired* chronicled was a *digital* revolution, of which the net and the web were mere expressions. Conveyed by code, digitization marked a paradigm shift whose promise and costs we still strain to weigh. There is a symmetry to my finally connecting with code through music, because music was where most of us first sensed the looming digital shift and its prickly tradeoffs, a full decade ahead of the web.

In musical terms the word "digital" is not synonymous with "electronic." The first known operational synthesizer was patented—naturally—by a Victorian, the American Thaddeus Cahill, in 1897. But the Mark II version of Cahill's *Telharmonium* weighed two hundred tons: there would be no Victorian synthpop. Experimentations in synthesis continued up to the 1980s, via the spooky-sounding theremin in the 1920s, the 1950s DIY imaginings of the BBC Radiophonic Workshop and artists like Delia Derbyshire (arranger of the groundbreaking theme from *Doctor Who*), and lesser-known independent pioneers like the Dutchman Tom Dissevelt, whose music from sixty years ago still sounds futuristic today. However, none of this music, not even the early work of synthpop pioneers like Depeche Mode or instruments like the Moog, was digital. Before the digital revolution, everything in our day-to-day lives reflected the mechanics of the universe itself by being *analog*.

As might be expected but is seldom said, the root of the word "analog" is "analogous." Grooves etched onto a vinyl disc could be used to vibrate a needle (or stylus), whose vibrations were converted to variable strength electrical signals *analogous* to the original sound. These analogous electrical signals would be sent through a wire to a loudspeaker, which would vibrate, reverting the signals to sound. In analog synthesizers like the Moog, voltage-controlled oscillators, filters and amplifiers generated the required electrical signals. This worked well, but analog signals could decay or be altered by circuitry between a turntable or synth and the speaker. Decay tended to happen at the edges of any signal, which is why analog synth sounds tend to be warmer and softer; less angular, sharp or percussive than sounds produced by acoustic instruments like pianos or snare drums or strings. In our digitized times it's easy to forget that the genius of the German electronic pioneers Kraftwerk consisted partly in their ability to coax previously unimaginable purity of timbre from analog machines. This was not easy.

Digital sound was a radical departure from analog, and its introduction previewed the revolution to come. Now a sonic signal could be converted to—or created as—a stream of *numbers*, just as Ada Lovelace had foreseen. And unlike an analog signal, a message composed of numbers is exactly (and infinitely) reproducible; will be precisely the same wherever it is sent, not to mention precisely manipulable. In contrast to other areas of our lives, where digitization would often be felt gradually, the effect on music was felt instantly because it could be *heard*. The first widely used digital synthesizer, the Yamaha DX7, appeared in 1983 and caused a sensation with the crystalline clarity and percussive *mass* of its sounds, as did the first digital sound processors and accessible samplers able to encode and reproduce real-world sounds. Suddenly recorded pop music acquired a sheen: it could glisten in a way never heard before, be glassier or more explosive, punchy or round, without the softening natural entropy to which an analog signal is by essence subject. Listen to any piece of electronic music from 1981–1982 and compare it to an equivalent from 1984–1985 to hear the digital switch come to life. Subscribers to a music streaming service can do this now: cue up The Human League's "Sound of the Crowd" (from the best-selling *Dare* album of 1981), followed by Scritti Politti's "Absolute" (from the equally popular *Cupid & Psyche 85*, 1985) to marvel at the difference in palette between two artists who sprang from the same post-punk electro foment. Or compare the original mix of Shannon's peerless "Let the Music Play" (1983) with the similarly fabulous "Buffalo Stance" by Neneh Cherry (1988), two songs doing more or less the same thing either side of the digital divide. And where music had gone, the rest of the world would follow.

Setting a pattern we have yet to come to terms with, musical digitization was received as a straightforward win at the time while proving more complicated over the long haul. The slender DX7 came with thirty-two preset sounds, some of which took musicians' breaths away, but it was excessively hard to program with its small digital display—meaning almost everyone stuck to the presets so records made or released in 1984 can be easy for a trained ear to spot. Most vivid of all DX7 presets was number twenty-six, "tubular bells"—most famously heard in the chorus of Band Aid's charity Christmas single "Do They Know It's Christmas?"—which appears on so many records made at that time that it might as well be a date stamp. Similarly, the new digital reverbs, noise gates and samplers had the democratizing effect of placing the sounds of huge rooms and orchestras and drums like thunder in the hands of everyone. Yet, shorn of the skill and determination—and money—previously required to create these sounds, they became ubiquitous and ultimately lost their value. Control and individuality had been traded for access and convenience. Not necessarily a bad trade, but a stark one that only became obvious when it was too late to reverse. This was a taste of things to come.

The computerist turned techno-sage Jaron Lanier finds a deeper portent in the digital switch, as we moved from recordings chiseled into acetate with vibrating ruby styluses to cascades of numbers encoded on a compact disc or streamed from a server. We dumped vinyl for three decades on the understanding that CDs reproduced a performance with digital precision. Which they do. So why are listeners of all ages returning to vinyl? Not just because, as the musician Jack White has it, "With vinyl you're on your knees . . . you watch the record spin and it's like you're sitting around a campfire—it's hypnotic." Nor because the album sleeve has yet to be bested as a platform for reefer production. Perverse as it might seem, the venerable twelve-inch plastic disc stores a lot more information. As Lanier points out, echoing the earlier argument about the limitations of PowerPoint slides, there is no space between numeric increments in a digital system. A straight line divided into nine segments provides ten points of quantization, zero through nine. In musical terms these increments define the limits of what may be expressed: nothing in between is possible because no "between" exists. By contrast, electrical or acoustic signals are graded, like inputs to the neurons in our brains, making the potential subdivisions between two points literally *infinite*—always amenable to further division—ensuring there is no limit to what may be represented analogously.

Lanier sees the stripping of nuance we first felt in music as a feature of most classical computing. Not necessarily because it's unavoidable. More because introducing nuance to the machines is hard.

"The binary character at the core of software engineering tends to reappear at higher levels," he writes in *You Are Not a Gadget*. "It is far easier to tell a program to run or not to run, for instance, than it is to tell it to sort-of run. In the same way, it is easier to set up a rigid representation of human relationships on digital networks."

Just as a logic gate is either open or closed, in the digital realm people are friends or not; single or not; happy, sad, angry, embarrassed or not. Complicated relationships and feelings, among other things, are always limited by the nearest increment; their essence is abstracted and approximated. According to Lanier, "[and] that reduction of life is what gets broadcast between friends all the time. What is communicated between people eventually becomes their truth. Relationships take on the troubles of software engineering."

Lanier's observations don't invalidate digitization, but they do articulate the limitations and circumscriptions we need to be alert to yet seldom are. The salient constant is that where upsides are usually foreseen, downsides tend to ambush us.

🐞 🐞 🐞

The venue folds into rail arches off an alley next to a construction site. Two performance spaces are connected by a maze of dark tunnels into which attendees vanish as if slipping through a vortex. Following the trail of bodies to the farthest space I find a young man and woman crouched behind laptops at a table, orchestrating a tsunami of glitchy, irregular beats that charge and chafe against each other in ways that make you both want to dance and wonder how the hell you would without several extra limbs. One large screen offers a stream of restless scratch images, another the code responsible for this novel sound, which is being written and rewritten before our eyes in an intimidatingly symbolic language I don't recognize. Do I like what I hear? I don't know. But I love the sense of experimentation and of artists pushing the envelope to see what works in this new medium. Later I watch TOPLAP's Alex McLean and two friends deliver a polished set of straight-ahead four-on-the-floor techno, while at the other end of the spectrum a woman named Emma Winston, appearing as Deerful, uses a minimalist language of dots, crosses, letters and integers called *ixi lang* to play and sing a shock set of sweet synthpop songs—the absolute last thing I expected to hear.

Amid a lot of promise, one act confirms for me that distinctive music can be made this way. Returning to the first performance space I see a young man with long blond hair crouched low to a laptop, hands arched over the keyboard as if intending to crawl through the display. At the first cirrus wisps of sound, his graphics screen fills with fluid black waveforms against a cobalt-blue backdrop, coiling and warping into each other in a hallucinogenic flow. As a crowd starts to gather and move toward him, variously textured washes of sound start to layer and shift, appearing to be improvised more like jazz than pop, except that you can *hear* the algorithms entwining in ways that would be hard to anticipate or predict. A few in the growing crowd try to move but most give up and close their eyes, letting the sound coil through them. This is lush and overwhelming, a new kind of music that is hard to imagine being made any other way.

The next morning I cycle under a warm sun to meet Alex McLean at Kings Cross station. On his way home to Sheffield and a little worse for wear after the excitement of the night before, he brightens over coffee as he traces live coding and its recent variant algorave back to the mid-1990s and a trailblazing language called SuperCollider (SC). As conceived by its American author James McCartney, SuperCollider consisted of two distinct elements: a sophisticated audio server capable of generating and/or processing any form of sound a user has the imagination and skill to find, and a language (called *sclang*) with which to control the server and form the sounds into patterns. McCartney claims his interest arose from exposure to the German *kosmische musik* fountainheads Tangerine Dream.

The sclang language was based on Smalltalk, fabled initiator of object-oriented programming in the 1970s, with syntax derived from the C family and sometimes compared to Lisp. Written in C++, the result looks more like Perl or JavaScript than Python, being multifaceted and rich but also convoluted and intimidating even to experienced newcomers, with brackets and semicolons littering the syntax—just the kind of thing I hate, or at least think I do. But at least I am not alone: faced with a precipitous learning curve, many contemporary live coders tiptoe around SuperCollider with a range of languages and frameworks built to address its groovy audio server in more accessible ways. Alex McLean has contributed an elegant member of this language cohort, *TidalCycles*, which is written in the sciencey Haskell, with myriad others including *Sonic Pi* (a program based on Ruby) and the *ixi lang* Emma Deerful used last night. Even a cursory glance at TOPLAP reveals a field growing apace. McLean regards live coding and algorave not as futuristic, but as an "almost Luddite" return to first principles, allowing abstractions to be shed in favor of direct engagement with the code most twenty-first-century sound boils down to but hides.

With a start I understand that all this richness of invention hauls me back to where I began: needing to choose a starter musical language based on principles and needs I can't yet see. To get a sense of SuperCollider's usability I go to see musician Shelley Knotts, who weaves and reweaves a spellbinding set out of nothing using a terse, highly abstracted form of sclang developed to ease the burden of typing during performances. In conversation I learn that Knotts, a student of classical music before she discovered SuperCollider, is on the purist side of live coding and algorave, believing there should be no cutting and pasting of code in performance. Others point out that languages like TidalCycles or the version of SuperCollider she uses in performance only work because they've been abstracted away from the SC server by lots of prewritten code stashed in black boxes we don't see—so what's the difference between that and cutting and pasting your own code? Another question still being addressed is how to manage the tension between concision and accessibility, because while no one wants live coding to be a measure of typing skills, excessive compression of syntax can reduce an audience's ability to follow a performance. How much does this matter? These details are still up for grabs. It's exciting.

The live-coding platform that catches my eye is Sonic Pi, a compact yet deceptively powerful offering built by a dynamic former Cambridge University researcher named Sam Aaron. By happy coincidence Nicholas knows Aaron from shared work on the Raspberry Pi minicomputer, so together we visit the Sonic Pi creator's Cambridge studio and come away impressed. Sonic Pi began life as a teaching tool for the Raspberry Pi, but its shrewd balance of functionality, immediacy and slick documentation allows a user to make listenable music fast while offering multiple ways to dive deep into the code. I launch myself at the platform, studying for a couple of months then setting myself the task of programming the extended instrumental intro to the New Order classic, "Blue Monday." And—a first for me and code—*doing it.*

CHAPTER

15

A Codemy of Errors

I can remember the exact instant when I realized that a large part of my life from then on was going to be spent in finding mistakes in my own programs.

—Maurice Wilkes, computing pioneer working on the world's second stored-program computer, EDSAC, at Cambridge University, England

There came a point when I felt I should explore other live coding options. I loved Sonic Pi, an astute platform with intriguing methods and quirky features and a friendly interface with jaunty colors that bely the imaginative complexity underneath—an overt invitation to make music in entirely new, algorithmic ways. What do I mean by *algorithmic* in this context? I can't point to any examples, but in my head I hear music directed by *conditionals*: if *a* happens then do *b*, else do *c*. Branching melodic patterns laced and interlaced using *for* loops and *while* loops and *elif* statements—shifting webs of sound, structured but threaded with controlled randomness, unlike anything achievable through traditional means. A kind of conditional jazz. Maybe *recursion jazz*, after the computing term for calling a function from inside itself, akin to plugging a video camera into a TV and then pointing it at the screen to create an infinitely receding series of images of a TV inside a TV inside a TV inside a TV . . .

I try to imagine what Philip Glass or Steve Reich or the electronic maestros Autechre could do with this new capability, feeling sure Bach or Miles Davis would have been equally entranced. And didn't John Cage spend most of his life designing algorithmic rule sets inviting chance into compositions that would compile differently every time? I don't know how to do this yet, but the possibilities *feel* endless.

I am starting to see the tradeoffs coders live with, though. The legibility of Sonic Pi's syntax comes at the expense of efficiency. My 32-bar intro to "Blue Monday" runs to 184 lines with comments, largely because of the way Sonic Pi sets the pitch and duration of notes with individual "play" and "sleep" keywords. This arrangement is easy to comprehend but means sequences of different notes occupy two lines per note, eating vertical space and making a long piece slow to navigate. The last half of Peter Hook's famous melodic lead bassline from "Blue Monday," for instance, looks like this in my rendering:

```
1 8.times do
2   play 38
3   sleep 0.5
4 end
5 play 38
6 sleep 1
7 play 36
8 sleep 1
9 play 38
10 sleep 2
```

Just as Ada Lovelace foretold, this code represents the MIDI note sequence 38-41-43-38, equivalent to low D, F, G and D on a piano. The first four lines show how Sonic Pi iterates, using a "do" keyword borrowed from Ruby. This four-line *block* translates as "do everything between the 'do' and 'end' keywords eight times": specifically, play eight low D notes separated by durations of half a beat (0.5), which is to say play eight notes in what musicians call "double time." The next six lines instruct Sonic Pi to play—or more accurately, instruct Sonic Pi to instruct the *SuperCollider audio server* to play—the notes F, G and D in durations of one beat, one beat and two beats, respectively. And here we see the issue. With six lines required to play three notes, a long piece can grow unwieldy for the fat-fingered naif, especially if—*when*—bugs emerge and they (*me!*) try to track them.

After scouring the documentation for a way to tie notes to their durations in a more concise way, I post my problem to the *in_thread* Sonic Pi forum and am tickled to receive a message from a man named Robin Newman, who astonishes me by linking to Sonic Pi code he's written for a piece by the sixteenth and seventeenth-century German composer Michael Praetorius—*really* not what I was expecting. Sure enough, this made extensive use of a "zip" method that ties corresponding elements from two lists (*arrays* in Ruby useage) to each other, in this case binding notes to intervals. *Exactly what I was after!*

Computing logic demanded that Newman's technique be deployed in three steps. These were (1) writing the zip method into a general function to zip arrays and execute their combined elements in sequence, (2) defining the arrays of notes and durations we want zipped and finally (3) passing our newly defined arrays to the function we created. Such a function might look as follows in Sonic Pi/Ruby:

```
1 define :playseq do |notes, durations|
2  notes.zip(durations).each do |n,d|
3   play n
```

```
4     sleep d
5   end
6 end
```

Ruby code is handsome, I think, with *goalposts* (straight vertical lines) rather than Pythonic parentheses used to introduce arguments/inputs in lines one and two. A detailed account of what's going on here is parked in the Notes & Sources, along with an example of an equivalent Python function, for comparison.

Now we can set up arrays for all the notes and durations in Peter Hook's lead bassline and pass them to our "playseq" zipping function. We do this within an iterative *do loop*, where we introduce the instrumental sound we want to use, an adapted piano sample I've called "hooky_bass," and then introduce some reverb to add atmosphere. This loop will cycle through each note-duration pair twice.

```
1 hookybass_notes = [38,38,38,38,38,38,38,38,38,36,3
2 8,41,41,41,41,41,41,41,41,41,43,38]
3 hookybass_durs = [0.5,0.5,0.5,0.5,0.5,0.5,0.5,0.5,1,
4 1,2,0.5,0.5,0.5,0.5,0.5,0.5,0.5,0.5,1,1,2]
```

```
1 2.times do
2    use_synth hooky_bass
3    with_fx :reverb, pre_mix: 0.8, mix: 0.6,
  room: 1 do
4      playseq(hookybass_notes, hookybass_durs)
5    end
6 end
```

A major improvement on my earlier forty-eight-line unzipped rendering of the lead bass melody. Better still, the genius of automation means I can use this playseq function to do the same for all the parts in *Blue Monday* . . . and for anything I write in the future.

🐞 🐞 🐞

I'm overjoyed at Newman's intervention, not least because, while I'm no expert on choral music from Praetorius's pre-Classical music era, something about its spare ethereality is modern-sounding and sympathetic to my pop-schooled ears. I tell Newman that Praetorius was a contemporary of one of my favorite composers, Gesualdo, an otherwise unhinged Italian count who experimented with musical languages much as we're doing with code, using scales and techniques—including dissonance—that would scarcely be heard again in Europe before the late nineteenth century, the opening of whose *Tenebrae Responsoria* comprises eight of the most beautiful bars of music I have ever heard. And so, to my utter delight, Newman sends me a Sonic Pi setting of a Gesualdo piece, writing:

"Glad you enjoyed the Praetorius. I enjoy setting any music of that era for Sonic Pi . . . I hadn't done any Gesualdo before so I thought I'd try one of his madrigals this afternoon. The resultant code is posted below. Enjoy!"

And in this moment I really do love all coders. There's a catch, though. Although based on Ruby, Sonic Pi is counted a DSL or *Domain Specific Language*, meaning a language used for one purpose within its own discrete coding environment. But the zip method is not part of it: Newman lifted zip from the underlying Ruby and said he couldn't promise it would remain stable under all circumstances—that it wouldn't "break" at some inopportune moment. I've stood on just enough stages to know that if a thing can go wrong in a live setting, it does, and I'm not sure I want to leave that possibility open.

But wait. I know from a conversation at PyCon that a man named Ryan Kirkbride, a doctoral student at Leicester University in the United Kingdom, has introduced a new live-coding platform called *FoxDot*. In common with competitors like Sonic Pi and TidalCycles, FoxDot allows users to bypass SuperCollider's gnarly SC3 language by offering a library of simpler methods with which to control the matchless *scsynth* audio server. Unlike the alternatives, FoxDot's methods are written in Python.

In the bloodless parlance of computing, what Kirkbride's platform does is "schedule musical events" much as sheet music does, but with algorithmic logic and the untranscribable magic of randomness baked in as options. The program will require plenty of new learning, but made easier by familiar Pythonic syntax. Better still, FoxDot is devilishly concise, with rhythms built using symbols to represent different percussive instruments (x, o,—, =), much like traditional percussion notation. In contrast to Sonic Pi, notes, their durations and envelope parameters are set in the same line by default: Kirkbride's entrancing demo on the FoxDot homepage shows how much can be accomplished in this way with a minimum of keystrokes. I defy anyone with even a passing interest in music to see this and not want to have a go.

As ever in coding, there are tradeoffs. A little musical knowledge is helpful with FoxDot and you wouldn't show it to a child or complete beginner in preference to Sonic Pi, because it looks like the kind of highly symbolic code Alan Turing might have used (this is the cost of concision). The genius of Sonic Pi is that you can get results quickly, with the transparent syntax and lively documentation making it fun to learn—which I suspect FoxDot won't be. I can't yet know whether the latter contains the worlds I saw in the former but am hungry to find out.

I set aside a couple of days to get started; feel like a sprinter in the blocks, impatient to set off. First step, install the program, the easy part, and then—

Oh . . . wait.

Here's how that goes.

SESSION 1

In a perfect world all programs would be written from scratch, with all necessary code integrated. But is every development team in need of location services going to rewrite Google Maps, when Maps is already available to use? No. During PyWeek I was introduced to the devilry of dependencies, beloved of programmers and enthusiasts of Murphy's Law, weavers of a rich and ever-evolving tapestry of ways to ruin a day.

What could go wrong? Multiple applications on a hard drive requiring the same dependency, but in different versions. A dependency modified or updated for one program turning toxic or invisible to another. Open-source developers' decision to stop maintaining an app on which others depend. But after facing these issues several times now, I think I know what to do. Python has a module called *venv*, which is used to create *virtual environments*. Venvs are a form of metaphorical container into which a project may be placed, along with its own private version of Python and whatever dependencies or packages are needed or desired. Once established, the virtual environment is Python's Las Vegas: what happens in the venv stays in the venv, protected from the coding badlands outside. Brilliant.

Naturally, I remember venvs *after* installing FoxDot et al. No problem: I can uninstall and start again. Except that when I go to uninstall, the craziest thing happens. I can't find FoxDot. *What the—?*

SESSION 2

A session spent playing cyber-Sherlock. Discovering at length that Apple, concerned for their users' spiritual growth, have embraced the character-building potential of teeth-grinding bewilderment by graciously hiding obsolete versions of Python 2 on all their laptops up to now. I know. We're welcome! Because I failed to specify "Python 3" instead of just "Python" during the install, Python 2 has risen like a kraken from the depths of my hard drive and dragged FoxDot under. Muttering oaths, I snorkel down to retrieve it.

SESSION 3

Deep breath. To create a venv for FoxDot I need the *command line*, accessed on Macs via a Utilities app called *Terminal* or on PCs by pressing The Windows key + x. The sense upon opening Terminal is of having cracked a hatch on your operating system and plunged inside. A small black box pops open. A line of text indicates what directory you're in and offers a dollar-sign prompt at which to type commands. That's all. Literally *all*. Nothing to point at, click, hear you scream. Two things I know about the command line are that (a) Real Coders love it because it strips away abstraction and allows them more control, like driving stick rather than automatic in a car, and (b) applications intended for Real Coders often *only* run and can be set up from it. I also know that while I've used it before, Terminal scares me. By the time I've found a decent glossary of commands for the laconic *Bash* command line language used to boss the machine ("ls" to list the files and folders in a directory, "cd" to move between directories and so on) another whole session is gone. I haven't *done* anything. But then I haven't destroyed anything, either. Call it a tie.

SESSION 4

The next day. Back at the screen. Ready for anything. Installation 2.0: bring it.

So. Download SuperCollider, optional "SC3" plugins and library of "quarks" used to connect to other programs, one of which is for FoxDot. Consult venv documentation and watch helpful YouTube tutorial. Open Terminal and type "python -m venv FoxDot_env" at the command line, creating a virtual environment with a directory I'm calling FoxDot_env. So far, so good. Now use the "pip install FoxDot" command to import FoxDot from among 450,000-plus prewritten packages of code in the Python Package Index, known colloquially as the Cheese Shop and instantly available from the command line via the "pip" command. Again all looks good. Until I fire up SuperCollider and then FoxDot, only to find they're not speaking. Oh no . . .

A terse error message ends "make sure FoxDot quark is running and try again." But the quark is running. Except . . . wait. My venv acts as a prophylactic to stop crosstalk with the outside. And SuperCollider is outside. *Eureka!* I bail on the command line and use Finder to drag SC into the venv, then close and restart everything. Still no dice. Lengthy consults with Google and Stack Overflow reveal that while venvs can be configured to allow resident packages to reach outside, in all cases they only work with *Python* packages. Which SuperCollider ain't. *Sigh.*

SESSION 5

Uninstall everything. Reinstall everything, *sans venv*. Find there's been a karmic bitflip overnight. Everything works! SC starts like an eager puppy. From a dropdown menu I install the FoxDot quark and recompile the Class Library as instructed. Follow the documentation in rebooting the server (did I do this before?) and type "FoxDot.start" into SuperCollider's IDE editor. Hit command+return . . . experience a rush of pure ecstatic joy as words and symbols slide easily up the screen, ending with the NASA-esque affirmation "Shared memory server interface initialized." SuperCollider is alive and listening for FoxDot!

I open a Terminal window, hold my breath and type "python3 -m FoxDot" as advised. Pulse quickens as more text appears . . . this time *not* ending in a warning. The two programs are communicating! I dare to go to the "Getting Started" section of the docs and follow instructions to create a basic synth player by entering p1>>pluck()—the parentheses *so* reassuring and Pythonic. To my amazement, FoxDot plays a synth sound called "pluck!" Tomorrow I can add a bassline and drums. The adventure begins . . .

SESSION 6

Take me now, Lord. Expecting to launch myself at FoxDot I'm thrown for another loop. And not in a good way. Seems that whatever terms I'd negotiated between FoxDot and SC programs have lapsed overnight. *How?!* Did they decide between themselves? I reload the FoxDot quark. No change. Close and open both main programs. Nuh-uh. Uninstall and reinstall, this time including my Python in the exchange. Nope. In each case I get the same long and cryptic SuperCollider error message ending in:

```
ERROR: There is a discrepancy.
numClassDeps 1116 gNumClasses 2230
```

Not to be outdone, FoxDot complains that it can't find the SC server.

The next half dozen thoughts zig-zag between hope and despair in a way I'm growing used to: dueling impulses to dig in versus walk away, scrolling the mind's eye like error messages sprung to life—

Don't get up, stick with it.
The carpet needs cleaning.
The carpet doesn't *need cleaning.*
God, Terminal's ugly
But I don't have any carpet!
This is too fucking hard.
I can do this.
(No. You can't.)
I'm hungry.
Did anyone feed the fish?
Wish I could be *a fish, just swimming around,*
never having to worry about SuperCollider.
I'll never be a goddam coder. . . .

And so on. Focus skittering like bugs in a dumpster.

Okay. Don't panic. You know what to do first: look to see if anyone has had this problem before.

I paste parts of the error message into search and am unsurprised to find that my problem has been encountered by others. If the internet has taught us one thing it's that nothing ever happens once. But I'm unnerved by the absence of solution, a lack duly repeated on the official SuperCollider forum. Thank Ada for Stack Overflow then: I can ask my own question there. Except someone already has. And none of the answers solve the problem. Despond gathering, I realize the only option left is the nuclear one, unleashed at the risk of blowing up my entire system. Hair standing on the back of my neck, I uninstall everything and resolve to come back tomorrow, when I will return to GitHub and try to install SuperCollider *from source*.

🐞 🐞 🐞

The feeling of installing from source, assembling and configuring a program from its disparate parts, is akin to being allowed out with the family car for the first time as a teenager. It strikes me that in coding terms I *am* a teenager, while so many real-life teenagers are raging adults. Is coding youth wasted on the coding young? Probably. What I am about to do is either smart or unconscionably stupid.

I try to pixelate away yesterday's dejection, hoarding hope the way a squirrel hoards nuts. Inspection of the SuperCollider source code repository on GitHub reveals lots of individual files and folders to be cloned onto my machine—the easy part—and a long "README" file describing myriad extra steps and what to do when they don't work, ending with "If all else fails, post to the user list stating what git hash you have checked out and all Xcode version and library information and most importantly the error messages . . ." A messy and mostly unintelligible (to me) reminder that this is open-source software, not commercial; that there is no helpline and no one to buttonhole for a refund when the code levee breaks, because everyone's working for free and by default you are as responsible for the product as they are.

The README file reiterates that SuperCollider consists of three parts: the SC3 language and interpreter; an IDE (Integrated Development Environment, basically a bespoke editor) in which one writes code; and the star of the show, the scsynth audio engine. An upgraded Supernova engine is also available to support the "multi-threading" some computer chips use to process complex tasks faster, but mine doesn't do that, so scsynth is good enough.

Scarier by far is the fact that installing and running SuperCollider involves a roster of requirements and dependencies as long as your arm, including the "command line tools" from Apple's giant and generally unloved Xcode IDE, access to the Homebrew package manager that facilitates orderly installation of software packages via the command line on Macs, and to git, which we know about—plus a trio of tools called "libsndfile," "readline," and a GUI (graphical user interface) development toolkit called Qt 5. The Cheese Shop will also come into play, along with my Atom editor and a program build management tool called CMake. This is the kind of complex chemistry that lurks behind most programs, but which most of us never see. It's a little different for Windows than my Mac, but not much. I'd had no idea. No wonder software goes rogue.

I think I already have pip, git and Homebrew on my machine, but open a Terminal window to check, successively typing "git" and "pip" and "brew info" into the command line. At each command text washes up the Terminal window as if sprayed from a hose, containing descriptions of the dependency programs, then notes on the dependencies' multifarious *own* dependencies and how effectively they're working, with instructions for fixing anything broken or upgrading to newer versions if available. None of which Finder or its Windows equivalent can do, so the command line starts to make sense: it provides valuable information. Lastly, inputting "brew doctor" draws a response of "Your system is ready to brew." Good to know. The degree of interconnectedness, of interdependency in this—and by extension most—programs is staggering.

I open Atom, the weird, modular editor I've recently taken a shine to, even if my affection took a hit when I found I couldn't use it with Ruby. Apart from the funky design, what I like about Atom is that it's open source and free and was developed to integrate easily with git and GitHub. After a bit of fumbling, I remember how to instruct it to clone the SuperCollider source code from its GitHub repo, and to my amazement it works. *Hurrah!* Hope buoyed, my attention turns to the requirements and dependencies as confidence soars. This is gonna to be great and I can't wait to—

Oh . . .

Wait . . .

🐞 🐞 🐞

There's a famous scene from *Dallas*, the Reagan-era TV melodrama about a Texas oil clan cursed with a rare genetic disposition to lip quivering, in which one of the show's few likable characters, Bobby Ewing, having expired under the wheels of a hit-and-run car driven by actor Patrick Duffy's movie ambitions at the end of season eight, rises from the narrative dead at the climax of season nine, ratings slipped and movie dreams forgotten. Selling this twist presented Dallas writers with perhaps the greatest storytelling challenge since the founding of the major religions ("Say yer holy book came from *where?*"): how to roll back an entire season's thirty-one episodes. Their now notorious solution was to have the whole of season nine dreamt by Bobby's wife Pam, who wakes to find her ex-late husband trying to scrub away the world's incredulity in the shower. Readers born into our present age of expanded credulity may strain to imagine the human race rolling its eyes in unison—may even already have some innovative theories about Hillary Clinton's role in all this—but that's what happened in those less siloed, largely algorithm-free pre-web days. All the same, my thoughts are with Pam because, two weeks after hitting my first *return* in the process of installing SuperCollider from source, I find myself waking as from a daymare, feeling much as she will have felt on spying Bobby's previously dead ass in her bathroom.

I don't propose to offer forensic detail, because this would fill an entire book and risk traumatizing readers of a sensitive or competent disposition. Suffice to say I uninstalled and reinstalled everything at least half a dozen times; spent eons hunting for faults in the requirements and dependencies, then in *their* requirements and dependencies—at which point my ordeal became, to borrow the coding word for a function called from within itself, *recursive*. I discovered Xcode and Qt 5 to be gargantuan programs from which I only required small parts; that only an old version of the latter was acceptable to SuperCollider (a fact mentioned toward the end of the documentation, when you'd already negotiated the licensing terms and installed the latest version of the behemoth). But this wasn't the main problem.

By the end of the first week, warnings and error messages pirouetted through my sleep, stalked by merciless command line orders to update, reconfigure, remove, reinstall, stand up, sit down and if you scroll further down, no doubt, don't slurp your soup or sit on the yellow snow—orders I'd spend entire sessions trying to understand and address, then more straining to action. In desperation I spent one whole session figuring out Docker, a kind of *über* venv that creates a standardized simulated computer (or *virtual machine*) within a user's local machine, allowing software to be delivered to a known, generic environment in which it will definitely work, every time, all unpredictability abstracted away. Cunning! *But* (I discovered far into the process of arranging a trial/installing/learning to use) prohibitively expensive for a non-pro.

And don't even get me started on CMake. I spent so much time trying to locate errant directories and files on my hard drive that I began to imagine myself in a digital remake of the famous scene from the rock mockumentary *This Is Spinal Tap*, in which the fictional band of the title almost misses a show after getting lost between dressing room and stage.

I admit: after a fortnight of my brain pinballing around the hard drive, I gave up, shamefaced, concluding Sonic Pi to be my lone viable option—knowing I needed to push on with the code rather than remain mired in the purgatory of getting to the point where I *could* code. Try as I did to stay buoyant, and much as I make light of my travails in retrospect, the wretchedness I felt at that point is hard to overstate. I can't recall the last time I felt so unconditionally defeated.

The following is going to sound made up or massaged, but I swear it's true. I had just reached the limits of what I'd previously imagined to be a substantial capacity for attrition, now exposed as pitiable hubris. The FoxDot problem had followed me from California sun to grizzly London gray via a book festival in Poland, betraying no inclination to shift . . . until finally it was too much. I shut my laptop and pushed it away. I remember a fingertip patter of drizzle on the window, a moped rasping by. Somewhere a siren. The code police coming to get me.

The next week I was due in Leicester to meet Ryan Kirkbride, FoxDot's modest and likable creator, who by now had responded to my plaintive cry for help by admitting he'd never seen these SuperCollider error messages before and didn't know what they meant. Sick to my stomach, I wondered how to tell him there was no point in us speaking now. That the thing I'd feared most from the outset—reaching some hitherto untested outer limit of my ability envelope—had occurred in the most unlikely of places . . . that I would be retreating to Sonic Pi with a humiliating new grasp of Sam Aaron's achievement in creating a rich live coding environment that worked *straight out of the box* every time. I had failed to mount the most basic step in making Kirkbride's work dance. What had I been thinking when I took on this miserable assignment? What an arrant, heedless, reckless fool.

I genuinely can't recall why I did what I did next. When someone later presses me for an explanation, I can't even trace a thought process behind the action. All I know is that, having given up and let go—having stood up ready to walk away—something impelled me to sit down, pull the laptop back, reopen SuperCollider and navigate to the Preferences section of the overhead menu in the IDE, which I had neglected in focusing on the command line. Maybe inspiration came from discovering in the course of myriad installations and reinstallations that, where there existed more than one version of a dependency—or *part thereof*—the host program could be confused as to which to use, with some choosing the most recent version, others applying some hidden logic to select one over another, still more simply freezing. Did the message "ERROR: duplicate Class found: 'FoxDot' . . . There is a discrepancy" imply such a conflict?

Preferences opened as a pop-up window with four tabs: "General," "Interpreter," "Editor" and "Shortcuts." I roamed awhile, before noticing something odd in the Interpreter tab. As we know, the interpreter translates the code I write into instructions for the SuperCollider audio server. I remembered that when we speak of "Python 2" versus "Python 3," what we're really referring to is not the language itself, but the specific version of *the interpreter*, each of which will be equipped with different capabilities, around which the language is arranged. Python 3, for instance, can process *f-strings* while 2 cannot. So, the interpreter is the nexus of everything. And there, under a line reading "Interpreter Options (stored in current active config file)," was a box containing:

```
"/Users/andrewsmith/Library/Application Support/SuperCollider/
downloaded-quarks/FoxDotQuark/FoxDot.sc"
"/Users/andrewsmith/Library/Application Support/
SuperCollider/downloaded-quarks/FoxDot/FoxDot.sc"
```

Just as specified in the error message. Two instances of the FoxDot quark? Had these confused the interpreter? I removed both and returned to the main menu, where I found a pair of quark options ready to be installed by checking a box. I chose one and rebooted the program, then fired up FoxDot. Nothing. Heart sinking all over again, I removed the first and selected the second option instead, rebooted and sat waiting for more error messages to creep up the screen . . . but SuperCollider started with much less ado, much less text—and *no error messages*. Happiness suffused me as the line "Welcome to SuperCollider" settled into place, with no red text in sight. Nervous now, I opened a Terminal and typed "python3 -m FoxDot," tensed in anticipation of the dread warning . . . which didn't come. FoxDot and SuperCollider were communicating.

Somehow, at some accursed moment in the installation process, I had managed to acquire a rogue quark without realizing it, and—this not being commercial software—there were no guardrails. That's all it had been: the metaphorical dash-that-should-have-been-an-underscore and took five minutes to fix but thousands to locate. I went back to Stack Overflow and answered the question no one else had, astonished as before that something so easy could be so hard. And as I finally set about exploring FoxDot I felt elation, but also grief. Two-plus lost weeks: think what I could have done with that time in a medium I understood.

Only the next day did I start to feel differently about my forfeited weeks. Through the process of tracking the stray quark and the antsy installation of SC from source, I got lots of practice with git and GitHub, the code banks through which most open-source software passes, while learning to run them from the Atom editor I no longer approached as if it were a skittish pet tarantula. I'd understood and learned to love Terminal and the command line, which seemed dour and gray before, but now bathed the machine in light and removed a significant barrier between myself and the coding firmament. I'd seen inside a complex computer program, was happy in the Cheese Shop and with Homebrew, and Python's handy venv module. I'd come to understand what Docker, CMake and Qt are; how requirements and dependencies work; how programs get built and where their pieces go on my system—also knowing various ways to find them when lost. I was the Time Lord of uninstallation. And reinstallation. And Uninstallation. And . . .

Later I went back to the PyGame games I hadn't been able to use and had scant trouble getting their multifarious dependencies to function, feeling mystified by my earlier difficulty. Somehow, searching for answers and even asking questions—so intimidating before—were becoming second nature, while programming suddenly seemed less about learning the obvious stuff like language syntax (important as this is) and more about negotiating relationships between elements of the microcosmic ambit. Perhaps the one true commonality between computers and people is that both effectively *are* their relationships, finding full expression only as part of a network. Some of our present troubles may stem from confusing what humans need from these relationships with what machines require—from reluctance to acknowledge that our requirements and dependencies are different and that projecting one set across the other diminishes both.

My ordeal has even furnished a new ambition, because now I know that the crepuscular "bash" files peppering my system are for the command line, which has its own rich and versatile Bash language and interpreter—together known as a "shell" and capable of controlling the machine on its own in sufficiently skilled hands. Bash duly joins Spanish and piano on one of the lists on my wishlist of to-do lists to wish to do.

The final and perhaps most important thing I'd understood was that all the positivity I felt at the end of my ordeal; all the chest-beating, triumphalist celebration of hope over despair that brought me to the giddy peak of being able to install a program on a computer *all by myself* could be crashed in a flash by a single bug or baroque set of documentation. At which point I would again curse every software developer who ever walked, skateboarded or scootered the earth. And that's okay.

CHAPTER

16

Do Algos Dream of Numeric Sheep?

An AI Suite

No one confessed the Machine was out of hand. Year by year it was served with increased efficiency and decreased intelligence. The better a man knew his own duties upon it, the less he understood the duties of his neighbor, and in all the world there was not one who understood the monster as a whole. Those master brains had perished. They had left full directions, it is true, and their successors had each of them mastered a portion of the directions. But Humanity, in its desire for comfort, overreached itself. It had exploited the riches of nature too far. Quietly and complacently, it was sinking into decadence, and progress had come to mean the progress of the Machine.

—E. M. Forster, *The Machine Stops* (1928)

PART I: MACHINE LEARNING DOES THE TWIST

It's easy to think we see algorithmic logic in life. Moving to California from New York as a five-year-old, I was transfixed by the sudden exposure to exotic creatures like lizards, snakes, frogs and newts, which seemed barely real, outlandish anachronisms from another age, another world. Sometimes on a sunny afternoon I still like to go sit in front of the house on a collection of railway ties used to mark parking spaces, where a vibrant community of bluebelly lizards scurry about their business as if *I'm* the anachronism—which in the scheme of things I might be. Sit still long enough and the reptiles either forget you're there or cease to care, at which point you might as well be one of them.

One day I set to wondering if it would be true to say their lives are governed by a simple lizard brain algorithm. The priorities of the dominant male, Fluffy, certainly look open to such interpretation. His hierarchy of responses to any new input, say another lizard or a strange man with a notebook encroaching on his patch (it's all the same to him as far as I can tell) would depend on the answers to a finite hierarchy of Booleans, namely:

Is it a threat?
Can I shag it?
Can I eat it?
Would I rather just sit in the sun anyway?

Or is this hopelessly algocentric? Put another way, is the universe more like us or a computer? Because if reality is a giant data-processing mechanism or algorithmic soup, as some computerists, far-out physicists and evolutionary biologists like Richard Dawkins suggest (its primary stuff being not particles but information), the implication would be that Ray Kurzweil and the Valley Singularitarians are right: that the thrust of our technology is toward artificially intelligent inorganic life that either absorbs or supersedes us, and that there's not much we can or should want to do about this. The cosmos and microcosmos will reveal themselves to have been one and the same all along.

At the end of 2019 I attend an annual weekend conference called Foo Camp. More accurately, Foo Camp is an "unconference," meaning the hundred or so invitees arrive on the O'Reilly Media campus in Northern Californian wine country and decide on their own collective agenda by signing up to run sessions based on their own expertise. Among an unusually racially diverse and gender-balanced international crowd are some colorful individuals, from ex-politicians and public policy wonks to leaders of startups and nonprofits geared toward humanizing tech. A contingent of high-level managers and engineers from Google, Facebook, Apple, Microsoft and IBM complements a heady collection of academic researchers and thinkers, making it easy to see how sparks could fly. As with PyCon and so much else, none of us know this will be the last Foo Camp for an indeterminate number of years, but it's a thrilling experience, like summer camp with footnotes: a timely reminder why, for all its faults, people want to be in this space.

A lot happens. Every discussion or debate I attend, from "Digitizing a Country" (run by an ex-president of Estonia) to an account of efforts to design better forms of social networking, to "Magic Mushrooms: The Next Goldrush" and "Black Mirror: Thinking about the Unintended Consequences of Technology," is challenging and inspiring. I steel myself to run or get involved in running three sessions, the most intimidating being my maiden code demo, of Sonic Pi, about which I remember little beyond my fingers turning to lead and the thought *OMG, this is a lot harder than the pros make it look*. Of more substantial note is a session I nominally co-run with a pair of serious machine learning professionals after deciding to combine the separate debates we had each proposed on AI. When one of our number, a young CS professor from the University of Chicago named Eric Jonas, insists on calling our discussion "AI Is a Lie!," I and the second expert member of our team, a creative ML developer named Carey Phelps, first recoil at the provocation. But Jonas insists. And he's right. What follows may be the most riveting discussion I have ever witnessed, in a room jammed with professors and engineers and researchers, for most of which I—as one of only two nonprofessionals present—resolve to shut up and watch as if it were theater.

Battle lines are quickly drawn, with the room evenly split between Jonas's and many of the academics' view of "AI" as ML dressed in the emperor's new clothes it bought off an Amazon Marketplace scammer—against often heated pushback from corporate boosters, especially Googlers, whose business model is built on the technology Jonas and others now question. But as the push-and-pull intensifies, something unexpected happens, as believers are forced to admit they know their "artificial intelligence" is not as sold, is not actually *intelligent*. Only when "the media" is blamed for this misrepresentation do I feel obliged to pipe up, pointing out that if we are going to use a word like "intelligent" we must account for how most people understand it. And most people understand the definition of that word to include the capacity for transfer learning, the ability to transfer knowledge between contexts. As already noted, present "AIs" are nowhere near this threshold and may take decades to approach it, may never get there at all. Thus the word "intelligent" as applied to "AI" is misleading, and this is not the fault of the media—the culprits are the people selling the technology. Neither is it okay for Google or Amazon to say, "Ah, but that's not what *we* mean by the word intelligent," any more than Facebook should be allowed to claim "Look, when we said 'friend,' what we meant was trolls and fascists and hackers in Macedonia . . . if you didn't get that it's your own fault."

The room goes silent when I say this and a truth suddenly dawns—that the argument in play is not about whether "AI" is being oversold, it's about whether active or passive overselling of the technology *is justified*. The tension is between Jonas's wing, who consider transparency paramount, and a pragmatic short-termist view that leveraging hype is not only permissible but necessary to unlock funding. By the latter account, venture capitalists and shareholders don't want to hear caveats or doubts or requests for patience, so they mustn't. The same can be true for academic institutions whose priorities increasingly reflect the Google grants underlying them. *Take the cash, do the work and leave doubts for later* runs this line, else the work won't get done. Fake it till you make it, in other words, it's easier to apologize than ask permission. The ends justify the means: Siren song of Elizabeth Holmesian Silicon Valley nihilo-capitalism.

Jonas and allies cite many instances of "AI" algos having written themselves to provide output their human trainers wanted to see, but for spurious reasons (a program designed to recognize tumors turned out to be recognizing rulers placed next to them, for example). Few of us here today would predict the extent to which the "it doesn't work" line falls away over the next several years, as machine learning techniques described as "AI" are refined. By the end of 2022, DeepMind will have used its "deep reinforcement learning" model to blow away a succession of very hard problems, including the creation of algorithms to predict how proteins fold and modulate a nuclear fusion reaction—breakthroughs of vast potential benefit to science and society. Who wouldn't applaud? And yet . . .

As debate rages I feel forced to confront something no one talks about: that ML/AI techniques like deep learning tend to rely on an in-built ends-related logic. We tell the machines what we want to see—say, a maximum number of points within a classic Atari arcade game (as per DeepMind's Agent57 algorithm)—and let them figure out how to get there. Which in a curious way mirrors the argument in the room about whether desirable ends can ever justify compromised or opaque means. Jonas (and most of the academic scientists I meet) consider the integrity of the means, the *process*, to be sacrosanct, because compromise there undermines results and trust over time. But the crux is this. If such ends-related logic has turned out to be less than sustainable for humans, as many of us have come to believe, why would we expect it to work better for machines designed to serve humans? The way we compute is changing. Machine learning algorithms make it possible for the first time to predict the outcome of processes such as the folding of proteins, but without doing the science to understand *why* they fold as they do. Does this matter? No one knows yet. The way we live and think is changing in ways that were not foreseen.

So begins my interest in Google. An engineer qualified to dangle that name at the end of their conference lanyard has an almost tangible aura. The company was founded by engineers whose faith in engineering is absolute: who collect coding Big Names the way fans collect celebrity selfies, a little breathlessly, while setting what reputes to be the highest bar of entry in engineering. Neither is this reputation absurd. Installing a couple of Nest security cameras before an extended research trip, I am struck by the degree to which ordinary hardware is animated, made *extra*ordinary, by software of such luminosity that it puts me in mind of the Victorian essayist Walter Pater's dictum that "All art constantly aspires towards the condition of music." For me the Nest software gets close to this condition.

An incongruity makes itself felt, though. Only after a couple of years living among the coders does it occur to me that I seldom meet Google engineers outside formal settings. There is always a contingent at conferences, often seeming to pass through like a school of exotic fish breaching the surface, both there and not there. And there are the Big Names—Guido, Andy Hertzfeld, and others—who did time. A gifted self-taught C engineer I grow close to through live coding works for a subsidiary. But Googlers are notable only for their absence at the Code for America brigades I join or Python meetups I attend. With tens of thousands of workers in the Bay Area, I would have expected to bump into them in the cafes, bars and venues of San Francisco and Oakland, or the South Bay tech hamlets of the Valley. But for the longest time I don't. After a while I wonder why.

For its first twenty years Google looked like a corporate hard fork; a new kind of company. Even when caught reconfiguring their search engine to the surveillance needs of the Chinese state and selling "AI" to the military under troubling enough circumstances that staff members resigned, the company shut these programs down where other tech big beasts did not. That enough Googlers felt willing and able to protest these lucrative activities was notable even in an industry where skilled workers are in demand, the more so because management listened. The same might be said of a global staff walkout following news that Andy Rubin, the former General Magic engineer behind the Android mobile operating system, had been awarded a ninety-million-dollar severance package after allegations of shocking sexual misconduct. Victory was limited in this case, amounting to a promise it wouldn't happen again. Nonetheless, it was hard to imagine workers at Microsoft, Amazon, Uber or even Apple organizing with such confidence.

Sixteen months after the Foo Camp AI debate, any temptation to view Google as a new model corporation imploded with a finality no one foresaw. Timnit Gebru was an AI ethicist, a star Stanford engineering graduate who chose her beat after early exposure to algorithmic biases. She was widely respected in her field and her presence at Google reassured scientists outside the company, implying awareness that its bottom line intersects others—that the straight lines and silos it trades in carry the potential for benefit *and* harm. All of which reassurance ended with a report on a then relatively new form of "AI" called the large language model (LLM), in which Google was developing strong commercial interest.

As the world now knows, LLMs work by training algorithms on vast datasets, typically harvested from the internet. Thereafter they work probabilistically, absorbing the rules of language by noting which words tend to go together in particular contexts. If the dataset is large enough, the algorithm gets good at mimicking human interaction via text or speech. The first widely recognized large language "AI," the GPT-3 system devised by the (then) nonprofit OpenAI, was convincing enough to fool people into thinking it was human with unprecedented consistency. But OpenAI's achievement came with dangers. It took little imagination to see how LLMs could be misused. When the early computer scientist Joseph Weizenbaum applied the disparagement "obsessive programmers" to the misfit hackers he encountered at MIT in the 1960s, who grew close to the machine and distant from people, they were only half of an equation he found concerning. Weizenbaum is celebrated for having created ELIZA, a mid-sixties program written to simulate conversation with a psychotherapist. As ever, the computer understood nothing but had been programmed to take cues from users' typed messages. Weizenbaum was alarmed to notice that however many times he told people ELIZA was *simulating* human communication, the appearance of meaning caused them to react as if it were real, as though the machine was sentient. Millennia of evolution, he realized, had left them unable not to. I would think of Weizenbaum in 2022, when a Google engineer publicly (and incorrectly) claimed the company's LaMDA "AI" had crossed the threshold into sentience.

I also thought about Weizenbaum while considering the word "empathy." Anyone familiar with the British Isles understands that while English contains dozens of terms for states of inebriation, such abundance does not extend to feelings, which is why it's no surprise that "empathy" is a recent acquisition, borrowed from the more conceptually expressive German—specifically the word *einfühlung* or "feeling into"—by nascent psychologists at the turn of the twentieth century. The original definition concerned not emotions passing between people, but feelings projected onto objects. Appreciation of a painting or full moon relied on an act of empathy, for example, and anyone who has even fleetingly felt their heart go out to a Roomba stuck under a chair has a sense of how powerful this empathic force can be. In his classic 1976 book *Computer Power and Human Reason*, Weizenbaum expressed concern at how this empathic force could be misused. Facebook's first president, Sean Parker, would duly later regret the way his company had harnessed "a vulnerability in human psychology" to its benefit.

Timnit Gebru and her collaborators cited Weizenbaum in their report on LLMs, which they called "On the Dangers of Stochastic Parrots: Can Language Models Be Too Large?" The authors' concerns were not limited to those inherited from Weizenbaum, however: they drew on a range of existing studies to flag a series of concerns, including the ocean of electricity it took to train LLMs effectively and the improbability of being able to screen giant datasets for bias or cruelty, or even analyze those biases once incorporated in a fluid algorithm. Moreover, because the technology was already being monetized, the direction of research was skewing to less socially beneficial uses.

The report was lucid and stylishly written but contained nothing new or controversial within the field. No mention was made of a Georgetown University study showing that GPT-3's efficiency improved in proportion to the extremity of input, because it hadn't been published yet. Few imagined Gebru would be fired in December, 2020, over a refusal to soften or withdraw the work, nor that her esteemed Ethical AI team co-leader Dr. Margaret Mitchell would share the same fate soon afterward, in February 2021. Thousands of staff and outsiders signed a petition of protest, while some Googlers quit, including one director who had been with the company for sixteen years and another senior employee who noted similar interference with a different paper on LLMs, whereby problems were glossed and mentions of Google's own systems removed. By mid-2021, *MIT Technology Review* was reporting that "Half a year later, research groups are still rejecting the company's funding, refuse to participate in its conference workshops, and employees are leaving in protest." At an SF Python meetup I attended that year, a UC Berkeley professor giving a workshop on approaches to summing numbers was careful to provide two sets of instruction, one for Pythonistas who use Google tools and another for refuseniks who will not.

In the dying days of November, 2022, the Gebru team's warnings acquired substance. OpenAI's ChatGPT LLM chatbot struck humanity like a comet, displaying a hitherto unapproached ability to simulate human intelligence through written text and speech. Within days educators realized their essay-based assessment methods had been blown away with no warning. Other early users noted the bot's tendency to parrot untruths and make stuff up; that purveyors of misinformation, disinformation, propaganda and spam had been gifted a mighty new tool likely to favor authoritarians and fascists; that intellectual property and copyright were suddenly more fraught. And a fresh range of white-collar jobs—including some forms of coding—were newly threatened by automation, offering a boost to inequality with no time for society to prepare. Most eye-opening of all were persistent public questions about whether ChatGPT and its "generative AI" successors, including Google's own Bard, were conscious, with many struggling to accept they were not. Mature experts had their work cut out trying to explain the truth.

By the time OpenAI's GPT-4 appeared in March, 2023, debate about LLMs had grown more measured. Educators had begun to see ways the bots could be helpful, and artists speculated about new forms of creativity, while debate over the technology was creating a well-informed public and perhaps attracting a generation of young people to science. Nonetheless, an impassioned open letter signed by an improbably broad range of names in tech, among them Elon Musk and members of Google's own DeepMind subsidiary, called for a six-month moratorium on new LLM releases in the interest of society and the species. Instead, an "AI" goldrush began.

To a casual observer Google's Ethical AI debacle looked like an abrupt pivot from "don't be evil" to the more Valley-normative "don't be evil and get caught." Or in terms a Perl coder might appreciate, DBE for **D**on't **B**e **E**vil became DBEUEBOCABAY (pronounced "debby eubocabay") for **D**on't **B**e **E**vil **U**nless **E**vil **B**uys **O**ur **C**EO **A** **B**ig-Ass **Y**acht. As I write in 2022, browser tabs on my laptop contain articles alleging a toxic, racist culture in Google's all-important site reliability group; on the company's prominence among a handful of industrial giants who promised to defund congressional objectors to the certification of the presidential election on January 6, 2021, then reneged; on the company's significant support for politicians behind nationwide voter suppression and abortion ban legislation; on it's embrace of the far-right demimonde built by Machievellian industrialists the Koch Brothers and the company's active and passive funding of global misinformation. Not to mention allegations, included in an antitrust lawsuit brought against Google by the Texas Attorney General, of a secret agreement (codenamed "Jedi Blue") to give Facebook "information, speed and other advantages" in Google Ads auctions as reward for backing off areas of competition. Tomorrow there will be more. As the day after.

At a glance, Google's recent actions look inconsistent with its founding principles. Some longtime staff certainly believe so, as attested by headlines like "Google Veterans: The Company Is Unrecognizable" (CNBC, January 3, 2020) and "'Techlash' Hits College Campuses" (*New York Times*, January 11, 2020). Added to this are the impressions of "insider ousiders" like the venture capitalist Martin Cassado of Andreessen Horowitz, who in July 2019 tweeted, "The brain drain at Google right now is astonishing. What the hell is going on over there . . . ?" But what if the company's new Machiavellianism is *not* inconsistent with its core mission?

The brilliant inventor Ray Kurzweil is AI's best-known theorist and advocate, thanks to a succession of books culminating in his 2005 tome *The Singularity Is Near*. If this sounds like an apocalyptic title, it is. The book is big but its thesis is simple: artificial intelligence in the true sense, incorporating consciousness, is inevitable. Once genuine AI manifests, intelligent machines will be recruited to design their successors, growing more powerful with each generation in an accelerating exponential rush to a kind of Big Bang for mind, a flash in which grounded biological life is transcended by pure disembodied consciousness—in which life becomes data.

The Singularity. Through which we transitional corporeal beings are either swept aside, become zoo exhibits or pets or make nice and get uploaded to whatever form of cosmic ether drive the machines choose to instantiate—but "as software, not hardware." Like gods we will be immortal, rapt, one, living in some cross between a computer game, a psychedelic trip and the lyrics to "Imagine." Whoever controls the technology will, at any rate. Parallels between the singularity and evangelical visions of the Rapture are hard to ignore—and the prophet Kurzweil is far from alone among technologists in looking forward to the future he describes. He even provides a timetable. Expect your toothbrush to become redundant in 2045.

Rather than ersatz religion, the sociologist Sherry Turkle thought she saw the hallmarks of ideology among AI activist-researchers she studied. As with the teachings of a Freud or Marx or Hayek or Rand, belief in the decisive power of computing—of *programmability*—created a prism through which all reality could be filtered and interpreted, providing believers with "a new way of understanding almost everything." Other researchers, like the anthropologist Robert Geraci, saw Kurzweil and the Singularitarians more conventionally, as logical heirs to apocalyptic Christian sects. The former *New Republic* editor Franklin Foer draws on Turkle and Geraci in his book *World without Mind: The Existential Threat of Big Tech*, seeking to build a case that Google founder Larry Page regards everything other than Singularitarian AI as incidental to his company's mission.

"When Page describes Google reshaping the future of humanity," Foer writes "this isn't simply a description of the convenience it provides; what it aims to redirect is the course of evolution, in the Darwinian sense of the word. It's not too grandiose to claim that they are attempting to create a superior species, a species that transcends our natural form."

If Foer is right and the point—at least for Page—is the machines, then Timnit Gebru was fired because she threatened Google's hidden core mission. Could such a thing be true? Impossible to know. What we do know is that in 2015, six years after helping set up Singularity University, Larry Page personally brought Ray Kurzweil to Google as director of engineering. The company's prohibition on nuanced discussion of its "AI" appeared to harden soon after.

Not everyone thinks Google's quest to produce true AI is realizable.

Back in the patent archives I found a predictable raft of entries for machine learning. Less expected was a collection under the general heading "computer systems based on biological models," generated by a vibrant niche community of researchers in a field known as unconventional computing (UC). Most of their work has yet to escape the lab, but its existence tells a tale much bigger than itself.

UC didn't spring from nowhere. For decades our expectations of technological advance rested on Moore's Law, an empirically grounded conjecture that the power of silicon central processing units (CPUs) could be expected to double every two years or so—which for the first half-century of computing proved accurate. I first heard time called on Moore in a meeting room at Intel at the turn of the century, from the lips of none other than the company's renowned Chief Technology Officer Justin Rattner. Fifteen years later at the Cambridge, UK, headquarters of the world's leading mobile chip maker, Arm, one of that company's founders, Mike Muller, reconfirmed industry fears that the transistors printed on silicon wafers to make classical computing chips were now separated by the width of *atoms*—were so tightly packed that quantum effects emerged if they drew any closer.

According to Muller, Moore was going the way of those laws preventing him and me, as British subjects, from eating swan or tethering our goats to Westminster Bridge after 2 P.M. on St. Swithin's Day. Did UC offer an alternative path forward?

Curious, I visited leaders of a well-funded EU-coordinated program of research into UC. I found scientists working with blobs of carbon nanotubes and tanks of bacteria. Others rocked slime mold, often seen growing on rotting logs and among whose roster of confounding superpowers is an ability to sense the shortest route between two points in a maze with no trial and error. (Could this be by some quantum sense? As I write, no one knows.) One researcher showed me how two chemicals, when mixed, formed themselves into exquisitely beautiful, perfectly distributed patterns, with precision supercomputers would take a century to match. The reason for this capability gap, he explained, was that classical computers could perform this task only by laboriously trying all possible solutions until the right one emerged—by "brute force" in computing parlance—much as a chimpanzee at a typewriter will eventually type a sentence that makes sense. This limitation stemmed in turn from the classical addressing method described earlier and which the Fortran pioneer John Backus dubbed "the von Neumann bottleneck," whereby we say to the machine, "Go to this (exact) memory location, do *this* (exact) thing with what you find there and put the (exact) result (exactly) *there*." As already noted, biology uses a template system of address, instructing "Do *this* with the next copy of *that* you encounter." Our adaptive immune systems present a good example of template-addressing in action: B cells and T cells lurk in the bloodstream until their "receptors" recognize a specific antigen target, at which point they multiply and go to work.

The main advantages of the biological addressing system are that a) it is error tolerant and forgiving precisely *because* it is imprecise, and b) it doesn't stand on sequence governed by a central processing unit: like the internet, analog biological systems are distributed, with many processes able to occur synchronously. Thus, where one compromised bit or misplaced comma in a digital program can crash the whole thing or produce unpredictable results, a damaged T cell has next to no impact on an immune response because another will come along soon. For this reason Alan Turing's colleague Jack Good preferred to call analog computing "continuous computing." Turing believed that a perfectly predictable and infallible machine could not be intelligent. The chief driver of intelligence for him was curiosity, so a true AI would need to make mistakes.

By the time I connect with Professor Martyn Amos of Northumbria University in the United Kingdom, whose 2008 book *Genesis Machines: The New Science of Biocomputing* introduced unconventional computing to the general public, aspects of the landscape have changed. First, a Dutch company called ASML has spent a reported nine billion dollars finding a way to continue cramming more transistors into chips, sparing Moore's immediate blushes, *again*. Second, machine learning models have followed search engines and social networks in signaling a shift to mimicking the statistical, probabilistic, pattern- rather than rule-based modes of nature exemplified by the brain or immune system—even if, importantly, the underlying substrate remains binary and digital. In response to these developments many UC researchers have parked the idea of creating non-silicon-based general purpose computers, refocusing on problems classical machines struggle with. But not everyone has abandoned the original quest, Amos notes, pointing me toward Professor Lee Cronin of the University of Glasgow, who also has pithy views on the present state of "AI." Not to mention Google.

"Google are librarians," he tells me in the kind of broad London accent you seldom hear in British academia and which usually betokens someone exceedingly bright. "They curate my reality for me in ways that I appreciate. That's all."

Beyond this curation, Cronin adds, Google is engaged in "bullshit." The universe does not compute, because it cannot be programmed for a predictable, reproducible outcome: people who think of nature or the cosmos as a data-processing mechanism ("that divine, Kurzweillian thing") are talking "crap." If analog computing exists, he says, it is only in the most limited sense. He doesn't believe consciousness or general intelligence can *ever* arise from classical computing and claims to have the calculations to prove it. His own faith is in "molecular" computing, using chemicals as substrate, but unlike many UC peers he believes in the continued necessity of abstraction—representing reality in the way of a classical "stack," using molecules instead of bits. He has already built a first working chemical computer and refers me to a fascinating paper on the novel programming language used to address it. And big players have shown interest: as we speak the US Defense Advanced Research Projects Agency (DARPA) has just added four million dollars to an impressive bank of research grants, specifically to build a "droplet computer."

Contrarian as Cronin's claims sound, I soon recognize them, and other researchers' efforts to compute using vibration, voltage, lazers and light—as concordant with ideas John von Neumann had been exploring at the end of his sadly attenuated life. Chapter one of an unfinished manuscript the great scientist left behind was entitled "Turing!" while chapter two was "Not Turing!" Among the Cronin Laboratory's big goals are (a) to figure out how intelligence works and understand its relationship to broader existence, and (b) to see if consciousness can emerge from synthetic life.

It doesn't take someone as smart as Cronin to know his goals will need patience, despite Silicon Valley's tendency to present artificial intelligence as either already here or soon to come bundled with Prime. Which hype is not new. Most accounts trace the modern pursuit of AI to 1956 and a small but august gathering of computer scientists including Claude Shannon (daddy of Boolean architecture), Marvin Minsky (legendary MIT AI Lab cofounder) and John McCarthy (creator of Lisp). The pre-meet agenda proclaimed an ambition "to find out how to make machines use language, form abstractions and concepts, solve kinds of problems now reserved for humans, and improve themselves"—with an estimate this would take ten experts a whopping two months to accomplish. We might ask why they didn't throw in a cure for cancer, demo of cold fusion and launch of a new teen dance craze. Yet failure in the fifties did not stop Minsky declaring in 1970 that "In from three to eight years, we will have machines with the general intelligence of an average human being," able "to read Shakespeare, grease a car, play office politics, tell a joke, have a fight." But if our first AIs were to resemble short-tempered car mechanics with a jones for amateur dramatics, Minsky didn't expect them to remain so, decreeing: "At that point the machine will begin to educate itself with fantastic speed. In a few months it will be at genius level, and a few months after that its powers will be incalculable."

And all before MTV could even arrive to reverse the gains at the start of the 1980s. How were Minsky et al. so cavalier in their assumptions? When I log into an electrifying online "AI" symposium hosted by the machine learning entrepreneur, academic and author Gary Marcus, I start to see the extent of the godfathers' hubris. Through the joy of watching Nobel laureates struggle to unmute their Zoom microphones emerges a startling revelation: even the people who know about this stuff know almost nothing about this stuff. The phrases most often heard from an impressively garlanded international panel are "We are still trying to understand . . ." and "We don't yet know . . ." One participant notes there is no agreed definition of what "computational intelligence" actually is, the remedying of which seems somehow *not optional* to the pursuit of computational intelligence. A rare hint of consensus clings to the idea that physical interaction with the material world (embodiment, in other words) may be a primary driver or even essential condition of intelligence, or at least would be if we knew what intelligence was. Which we don't.

Panelists veer between tentative hopes for new avenues to AI and multiple mystifications at the current direction of travel, the latter including (a) the present obsession with words and LLMs "when so much understanding comes in other ways," compounded by research suggesting that, contrary to decades of assumption, language is not essential to complex thought; (b) attempts to emulate the brain, whose properties are "just radically different from a manufactured object" and probably won't be understood for "a century or two"; and (c) any hope that intelligence could emerge from the "neural networks" on which "deep learning" models depend, which is "just wrong." The vastly experienced UCLA computer scientist and philosopher Judea Pearl echoes objections I've heard previously in declaring that data alone will never lead to intelligence, because "data is a window, not a foundation on which intelligence could be built." Summing up the current state of play, Pearl adds that "it is very hard to find a needle in a haystack, but it's especially hard if you've never seen a needle before." Host and co-organizer Gary Marcus points out that the word "deep," as in "learning" or "mind," has become a marketing concept used to suggest "cutting edge" to investors, the modern heir to appending "-omatic," "-onics" or ".com" to brand names in the fifties, sixties or nineties. In the wake of ChatGPT the same will be true of the signifier "AI."

At the risk of sounding perverse, I find the discussion uplifting. More than five hundred people are watching around the world. Good academic scientists are on this: they've got our backs, edging forward but raising flags as they go. Whatever "AI" means and however long it takes to arrive—even if it never does—it is encouraging a more nuanced appreciation of the miraculous complexity of evolved life. And the prime beneficiaries of all this research, of the dead ends and blind alleys attending all pursuit of knowledge, might be members of the nonhuman animal kingdom, whose consciousness and particular forms of intelligence are suddenly recognized as special. It's no coincidence that the cognitive abilities of corvids (the bird group including ravens and crows), cephalopods (octopuses, squid, cuttlefish), rays and even bees are now being viewed with respect and awe. We owe this new regard in large part to AI research. After repeat failed attempts to create computer vision algorithms, the machine learning specialist-turned-neuroscientist David Beniaguev and his team conducted a detailed analysis of human neurons, the brain's equivalent of switches in classical computing, concluding that the most powerful human-made "deep neural networks" would struggle to attain the sophistication of a single neuron; that each of a neuron's dendrites, the tendril-like structures that receive input, may in themselves be carrying out processing tasks beyond the capabilities of any current electrical system. If true, each of our brain's ten billion neurons amounts to "a sophisticated computational unit" on its own and the brain's potency is far greater than previously imagined. Knowledge such as this is humbling in ways that can only be helpful. Montreal panelist Daniel Kahneman acknowledges as much.

"I first became interested in the new AI when I talked to [DeepMind founder] Demis Hassabis," the Nobel laureate psychologist says, "and I am really struck by how modest he appears to be now compared to six or seven years ago."

Perhaps we are growing up as a species. And yet nothing can disguise the anxiety many of these AI movers and shakers feel right now. A few short years ago the most searching concerns about "AI" sprang from outside the industry, but not anymore. Algorithmic bias in present machine learning is raised repeatedly by members of the panel. Professor Celeste Kidd, whose UC Berkeley lab examines the formation of beliefs, mentions Timnit Gebru in declaring, "Right now is a terrifying time in AI. What Timnit experienced at Google is the norm: hearing about it is what's unusual." Gebru's former Google Ethical AI team colleague Dr. Margaret Mitchell is present for the discussion and describes the many stages involved in training ML algorithms, in which the values and assumptions of trainers cannot but intrude—and the casualness with which this fact is treated in the Valley. Tech leaders, she points out, are good at selling the benefits of their work but poor at considering long-term risks and dangers. The scramble to deploy LLMs post-ChatGPT will confirm her view.

"You wouldn't think it would be that radical to put foresight into a development process," she says, "but within tech it is."

Google and others have now introduced principles for the development of AI, but Ryan Calo, a professor at the University of Washington School of Law, expresses skepticism at these gestures, frowning:

"I have a few problems with principles. Principles are not self-enforcing, there are no penalties attached to breaking them. My view is principles are largely meaningless because in practice they're designed to make claims no one disputes. Does anyone think AI should be *un*safe?"

Calo thinks we need to "roll up our sleeves" and change laws to account for the range of new social complications brought by ML/AI code. Summing up the discussion, moderator Gary Marcus reflects the concerns of many researchers present when he says, "Pandora's box is now open. [This is] the worst period of AI. We have these systems that are slaves to data without knowledge."

PART II: LOST IN THE GOOGLEPLEX

Eventually, inevitably, code gravity does pull me to Google. Timnit Gebru hasn't yet been fired. The militarist Project Maven and Dragonfly Chinese search engine have been exposed and protested by significant numbers of staff—then stopped. I've met for a drink and spoken subsequently with Jack Poulson, the fiercely bright computational scientist and Dragonfly resister who has been working to flag Big Code dissemblings ever since. The fragmentation of America touches Google too, he told me, through the thousands of pulsating noticeboards and constant debate that make it so intriguing an entity, an ersatz global beachhead in the culture wars. *Wired* would later report that Larry Page's cofounder, Sergei Brin, had been "extremely reluctant" to enter China; he had even argued to withdraw after a state-sponsored cyberattack targeting human rights activists. Company engineers implored managers to drop Project Maven from the start, it seems. All of which makes me feel well disposed toward Google, certainly more than Amazon or the social media cuckoos in our nest, whose fealty to the communities they claim to serve appears feeble at best. Google may be tempted by evil, but for the moment their embrace of it looks negotiable. And we know they care, because after the Dragonfly fiasco they published a set of principles.

So I'm feeling open and light as I roll into a parking lot on Google's vast Mountain View campus, fresh from my visit to the nearby and unexpectedly fabulous Computer History Museum, still chuckling at Cray Research Inc.'s kink for surrounding its cool-as-Kubrick seventies supercomputers with bench seats upholstered in funky red leather, as if Bianca Jagger might drop by en route home from Studio 54 and need somewhere to sit while her program runs.

I've come to spend the afternoon with one of the company's top Python people, to talk about Python, programming, Google, and how code transmits to the beyond. At least that's what I've imagined.

Google's Mountain View campus is called the Googleplex and has the footprint and population of a small town. Modeled on a university campus because by admission that was all Page and Brin knew when it was built, its pleasant green sprawl has become the default model for dystopian tech-fi like Dave Eggers's novels *The Circle* and *The Every* and Alex Garland's Hulu drama *Devs*, where it takes on a hallucinatory flatness evocative of William Goldman's adapted satirical horror movie masterpiece *The Stepford Wives* (1975). As so often in life, the reality is at once weirder and more banal.

It's a bright, clear day and sun streams through the vaulted reception area in one of a seemingly infinite parade of low-rise, Lego-like glass and steel boxes as I present my credentials. There are no nondisclosure agreements or searches or mindmelds: security was heavier for a Python meetup at Yelp in the city. The vibe is collegial and easygoing. My welcoming host arrives and we take a tour of the building, which true to legend consists mostly of offices, workspaces, meeting rooms, and airy kitchens stuffed with every snack known to humanity, each qualified to double as a Noah's Arc of Fast Food should the floods ever return, majoring in those we craved as kids, as if architected by Hansel and Gretel. One of the few certainties I'll leave campus with is that no Googler ever went hungry.

I've been looking forward to this. To me, and I think to Guido, you can't talk about Python—or any other language—without reference to the environments it's written and deployed in, so I expect a ranging conversation attuned to gray areas and big pictures. If Google isn't about those things, what is? Yet the discourse remains narrow as twine and unshakably technical, my attempts at expansion ignored or talked over. When I raise the Project Maven and Dragonfly spats, presuming my ground safe after what was sold as a kind of corporate ethics New Deal, I have the strongest impression that my interlocutor doesn't know what I'm talking about. Either that or the issue is of no interest because responsibility ends when the code leaves the editor. I wind up back in the parking lot feeling like a kid who's been playing pin the tail on the donkey for four hours, disappointment and boredom compounded by shame at these states, not least because I thought I was incapable of boredom.

I drive to another part of campus, where a Googler I grew friendly with at a conference has invited me to swing by for beer. What I know about him is what I learned that weekend. A very senior engineer, he is also an accomplished classical musician with an original mind and bone dry wit skewing to offbeat—and often very funny—analogies between computing processes and human ones. At the conference we cleaved to matters of obvious shared interest, like code, music, science, the book I'm trying to write and a private code project of his, but we did stray onto families and personal stuff. I enjoyed the way he talked about his kids, and even more so his wife, with real love and emotional intelligence; wisdom. I learned that he wasn't a Google lifer, having moved west to take up his post only a few years previously, but I sensed—and understood—his pride at being there. I could see he liked his life in a rented pile in a Mountain View subdivision full of other highly paid company managers, while insisting he would never buy in the Bay Area "at these crazy prices." In common with many of my more established friends, he was mature and worked hard, but knew how to park the maturity and have fun. Given that he had devoted his work life to code, where mine had been given to the probably pointless task of trying to figure out people, I assumed we would think differently about some things, but I couldn't remember the last time that had been a problem with anyone outside my own family.

This is the first time we've met away from the suspended reality of a conference, and we're pleased to see each other. We wander onto campus and snap some pictures with the famous tyrannosaur skeleton outside Building 43, then enter my friend's office, which feels markedly different from the buildings I entered earlier, being a place where the code sausage is actually made and the public seldom goes. Again the first impression is of air, space, light; of high ceilings and acres of glass, nothing like a cost-conscious traditional office. Only when I draw close to desks and get into people's actual workspaces, full of knick-knacks and photos and conference swag, does it start to look familiar, affirming that some things endure even here.

We stroll to the edge of campus, to a homely spit-and-sawdust joint next to Highway 101. Pitched between sports bar and backyard spring break bash in atmosphere, the place is heaving with

Googlers kicking back at the end of a day's work or sharing a beer ready to return for more. The crowd can't possibly be as young as it looks, but otherwise reflects the demographic balance at the company, which means double the number of men to women and not a Black person in sight—balances claimed to be improving at Google, one of the few tech firms to publish figures, but s . . . l . . . o . . . w . . . l . . . y. While my friend fetches drinks I scope for precious seating, settling on a long table at the back of the yard, where a gap has opened between a small knot of young men and another lone male nursing a beer on his own, despite not looking old enough to drink alcohol in California. From an abundance of politeness I ask if he'd mind my taking the free space, but quickly see he is stuttering and struggling to form sentences. Accustomed to the neurodiversity of my new environment—and easily sold on its benefits—I smile and answer my own question, invite him to join us if he'd like but am not surprised when he opts to stay where he is. He may be working through a code problem.

My friend arrives with the beers and we chat happily, as before, first about my earlier meeting with the Pythonista (I am vague, not wanting to be a downer), then the conference where we met; who we've kept in touch with; my progress with Python, Sonic Pi and FoxDot; our home lives and respective impressions of the Bay Area as relative newcomers . . . which is how we stray onto the subject of the region's seething homelessness problem.

Things to know. You've been to western cities and think you've seen homelessness in affluent places. Nuh-uh. Until you've been to LA's Skid Row or the Tenderloin in San Francisco you have not. Every time I traverse the latter or dodge down Mission Street to some coding office or event, my silent prayers align with most visitors': please help Californians decide this parade of misery is worth spending money to fix *and please don't ever let me get used to what I'm seeing*. Worse, officials say up to 80 percent of the Bay Area's problem is hidden by an expanding population of individuals and families living out of cars or trailers, because tech-inflated rents are so high and scant safety nets so porous that almost anyone can fall through the cracks. At a party to honor a new digital art space in San Francisco, thrown in the opulent townhouse of a Big Tech marketing exec, I don't know whether to laugh or cry on being introduced to the institution's creative director and finding him to be among this invisible army, couch-surfing with the few other non-tech creatives still clinging to the gilded ledge.

I know my friend takes pride in a radical contrarianism common to a subset of coders, who by necessity are always looking for unconventional ways to solve technical problems. But that doesn't prepare me for what comes next, as he explains with patient reason, as one might to a child, that he thinks I worry too much about this; that he doesn't, because he figures millions of people want to come to this country, all the time, and they continue to want to come because everyone knows there are opportunities here (and that this must be true, or the market would speak and they would stop wanting to come). So if the opportunities are there and people fail to make use of them, well . . .

He shrugs.

I try not to stare, but I'm reeling. If people are on the street, it's their fault. All of them, equally, with no qualification or distinction allowed for circumstances. Fresh from researching George Boole, I recognize this argument as near identical to the one advanced by Sir Charles Trevelyan during the Irish Great Famine of 1845–1852, that homelessness and starvation were the natural consequences of a feckless, indolent population and Irish working people's plight sprang from a failure of character: here was evolution in action, in the form of God's own will. And I don't know where to begin in trying to empathize with this view enough to be able to address it—am not sure I even want to. We change the subject and carry on talking, but for me the conversation never recovers the ease it had before. We part on friendly terms, but I go away feeling wretched.

On the hour-long drive from south Silicon Valley to the bucolic North Bay I try to parse what just happened. I'm shocked by what my friend said, but more than that I am shocked by my shock. Such pitiless views are not unfamiliar to me: I know and even have people in my extended family who hold them. The novelty for me is hearing them from someone I like. How can such a callous—and factually indefensible—analysis exist within a person who does not strike me as natively callous or cruel, who is highly educated and versed in critical thinking? Plainly it can, and does, which means the source of this dissonance is internal to me. And as darkness starts to fall with the Golden Gate Bridge looming on the other side of the city, an answer floats to the surface. The key to my upset, to the disturbance I feel, is—how had I not seen this before?—*abstraction*.

I understood my friend's perspective to be characteristic of a libertarian. Like eugenics, to which it bears more than a passing conceptual resemblance, libertarianism didn't originate in California but found a fertile breeding ground here. Anyone who has read this far will already have enumerated the ways in which my friend's argument falls apart on even glancing contact with actual life as it is lived. The few other noncoding issues he and I moused over yielded a similarly shallow grasp of detail. But for him this didn't matter. His abstracted conceptual framework rendered particularities moot. Any appeal to empathy or compassion was viewed as lacking intellectual rigor; sweet but emotional and naive, these latter qualities amounting to the same thing.

Like anyone in the Bay Area I've encountered this mindset before. The libertarian house journal is called *Reason*, which is at once pretentious and enjoyably stupid, given that experience and an expanding body of science show that reason and emotion work together in a healthy person (and are equally reliable commissioners of action.) Over two entertaining conversations with the Princeton philosopher coder and AI researcher Tom Griffiths, author of *Algorithms to Live By: The Computer Science of Human Decisions*, I am provided many examples of this symbiotic relationship. An ability to love irrationally, for instance, is entirely rational, being essential to making someone a reliable life partner or parent. Indeed, a lack of such irrationality is common to what psychologists call the "dark tetrad" of personality traits—psychopathy/sociopathy, narcissism, Machiavellianism and sadism—a constantly evolving category believed to touch a substantial proportion of the population at a nonpathological, subclinical level and anywhere up to 5 percent in diagnosable clinical form. When not touting the imminence of machine consciousness, Marvin Minsky pointed out that in a normal person even anger is rational, shutting down empathy and allowing us to confront the snarling tiger free of an urge to muse, "Yes, but it probably had a tough upbringing."

I pull into my little hippy enclave grateful for the Tibetan knit shops and yoga studios, and even (sigh) the freshly opened psychic's studio, understanding why I was upset. My friend was speaking of people *as* abstractions, like variables being passed to a function, shorn of experiential context in a way I found distressing. In fact, *frightening*.

Am I being oversensitive? Maybe . . .

But my friend said something else that caught my ear. Curious as to whether the software engineering bar really was higher at Google, I asked how his department compared to other coding workplaces. Was there more rigor?

"I don't know," he said, smiling and shaking his head. "But there's definitely more crazy."

I wondered what he meant.

"Well, you should see what happens when there's a disagreement. It goes from zero to a hundred in seconds, with everyone shouting at each other—it's insane."

As he spoke, I understood he was describing the traits of autism, particularly those associated with the profile for Asperger's syndrome, a fascinating condition associated with intense and narrow focus, the love of repetition and control, and an almost supernatural gift for discerning patterns and retaining information obscure to more neurotypical peers. The cause of my friend's amusement was the fact that people with these traits typically struggle to read social cues, the subtle ticks of voice, face and body most of us use to intuit (and respond appropriately to) the thoughts and feelings of others. For many autistic people this is a source of great discomfort, which the comedian Hannah Gadsby, who is herself autistic, explains well when she says:

"You know how sometimes you put your hand under running water and for a brief moment you don't know if it is hot or cold? That is every minute of my life. Being perpetually potentially unsafe is a great recipe for anxiety."

I asked my friend if he thought Asperger's conferred some advantage.

"I guess it must," he replied with a noncommittal shrug, "because almost everyone on my team has it."

I order some books and reach out to Steve Silberman, San Franciscan author of the superb *NeuroTribes: The Legacy of Autism and the Future of Neurodiversity*. Since his elegant and humane book was published in 2015, Silberman tells me, our understanding of autism has been evolving at pace. Most of us are familiar with the idea of a "spectrum" running—in order of relative impact—through "low-functioning" and "high-functioning" autism to Asperger's syndrome. The problem with the spectrum metaphor is that no two autistic people present with quite the same combination of recognized autistic traits and that these traits are also widespread in the general population, where they tend to read as (often very useful) quirks. For this reason, the author explains, autism is increasingly conceptualized as a cloud of traits concentrated in, but not unique to, people diagnosable with the condition. Asperger's is no longer recognized as a diagnosis distinct from other states of autism, despite the term still being used informally and by organizations like the Asperger/Autism Network and Asperger's Syndrome Foundation at the time of this writing. Increasingly, "autistic" and other atypical traits—and almost everyone displays some of these—are folded into the term *neurodiverse*, as implied in the title of Steve Silberman's book.

In a 2020 book called *The Pattern Seekers: How Autism Drives Human Invention*, Sir Simon Baron-Cohen, director of the Autism Research Centre at the University of Cambridge (and yes, cousin of actor Sacha), claims the genes associated with autism are ancient, widely distributed, and important to our creative development as a species. Only when concentrated in individuals do they become problematic, he says, describing a careful experiment in which he came close to establishing that people with isolated autistic traits are far more likely than average to produce autistic offspring. Baron-Cohen made a splash early in his career by identifying autism with what he called "the extreme male brain," although he was careful to stress that the genderization he proposed was statistical rather than absolute (by his assessment my own brain is highly "female," for instance).

Whether genderizing autism is helpful or not, we can see where the professor was coming from, because while the Centers for Disease Control and Prevention (CDC) estimates that one in forty-four US children is diagnosable as autistic, it is conventionally held to be four times more common in boys than girls. The equivalent imbalance is often given as more extreme for an Asperger's profile, although female socialization can make the condition harder to detect in women, complicating the figures. Nonetheless, in his book *The Essential Difference*, written before the changes in diagnostic approach, Baron-Cohen claims that in people then diagnosed with high-functioning autism or Asperger's syndrome, the ratio "is at least ten males to every female." He also states flatly that "Autism is an empathy disorder," while being quick to point out that this does not equate to being cruel or uncaring, simply to having trouble seeing the world through other people's eyes—a trait he has called "mindblindness."

The point is this: autism and what used to be diagnosed as Asperger's syndrome are common among humans, especially male ones. While these conditions may confer some advantages in coding tasks like debugging, I have seen no evidence that they are markedly more widespread among the general coding population than elsewhere. And yet my friend charged that they were dominant among his colleagues at Google. What did this mean? Could it also bear on the jangling disjunction between his and my modes of thinking?

PART III: FORMS OF ABSTRACTION

When I flew to Germany to have my code naif's brain scanned for the first time, it was under the aegis of a groundbreaking study into how the brain deals with code. And although this was a new field of inquiry, the results looked momentous. Among a cohort of mostly computer science undergrads at the University of

Magdeburg, the main cerebral regions activated for code comprehension were those associated with natural language processing, located in the detail-oriented, analysis-loving left hemisphere. My scans were anomalous in showing significant activation in both hemispheres, but this was easily ascribable to a newbie's inefficiency at managing novel input as compared to an expert. The hamster turning the wheel up there had to sling a lot of mud to hit the right spots, if you will. Dr. Siegmund guessed future scans would find the luckless creature doing better due to increased experience, although these may not have been the exact words she used.

As a newbie the German study made sense to me. Most of the things I found hard to think my way into were finicky details, immutable rules, rigid binary logic. Yet, as I began to move beyond enervating spats with syntax and logic, communication with the machines began to *feel* less binary and programs less linear as they meshed with myriad others locally and around the web—and of course those killjoys who ruin everything, *users*. Narrow-frame, binary thinking looked necessary to many specific tasks but woefully deficient and even dangerous in the round. Much as heat changes the fundamental chemical structure of an egg, so code seemed to open out into something more fluid and elliptical the further you ventured into it. For me this came to represent its beauty and its challenge.

The Magdeburg team and I stayed in touch and had begun to discuss the planned follow-up scan when a pandemic intervened and forced a long hold . . . then longer . . . and longer. Until, just as the global crisis began to wane and travel was again thinkable, an email arrived from Nicholas that turned the code comprehension debate on its head. From the beginning my friend hadn't wanted to know too much about how brains process code, as if awareness would constitute a loss of cognitive innocence, like catching sight of the Tooth Fairy as a kid. But even he recognized the link he now sent to be eye-popping, saying, "I thought of you when I saw this." I clicked the link and was thunderstruck.

An MIT team had conducted a study similar to the first but appearing to draw polar opposite conclusions. What's more, the team in question had been overseen by someone central to our understanding of language processing in the brain, the redoubtable Russian-born American cognitive neuroscientist Evelina Fedorenko—whose name I knew from a captivating *New Yorker* piece on hyperpolyglots, those rare people for whom language learning is like breathing. We exchanged pleasant emails and she put me in touch with the code comprehension study's leader, a young cognitive scientist named Anna Ivanova, who walked me through the similarities and differences between the two studies, being careful not to criticize or overclaim. Only toward the end of our conversation, when I asked directly, did she confirm that "Yes, our findings do contradict theirs." We agreed on a visit for a scan and more detailed discussion as soon as the lab reopened to outsiders. At the end of November 2021, it did.

MIT's Department of Brain and Cognitive Sciences is set among a cluster of technical blocks in the cosmopolitan east of Cambridge, Massachusetts, and must be one of the most stimulating places on Earth to work right now. Over the course of this project, the curiosity of scientists has been a precious counterweight to the bluster and opportunism endemic to the business side of Silicon Valley—extremes of impulse between which coders are often caught. And so it is here. Known affectionately to members as "EvLab," the Fedorenko Language Lab is located off an airy central atrium around which other specialist labs cluster, with the all-important fMRI scanner on the ground floor below. I'll spend two days with young EvLab scientists and the multidisciplinary team behind the code comprehension study, to leave feeling renewed and optimistic, buoyed by the infectious joy of discovery. Over pizza with the team on the final night, the brilliant Indian lead coder Shash Srikant teaches me hash functions and helps me think through a problem with my Shakespearean Insult Engine. Jan joins us and is similarly enchanted. I can't imagine being happier. What I see around the table is a vision of our future as rich and sustainable.

To business. The MIT study "Comprehension of Computer Code Relies Primarily on Domain-General Executive Brain Regions," titled for the Radiohead song of the same name (kidding), had the same aims as its German predecessor, but with important differences in methodology. Subjects were fed code snippets in either Python or the graphical educational language Scratch, rather than Java. Around 60 percent of subjects were women and a majority natural scientists with good coding skills, while the Magdeburg cohort had been composed of overwhelmingly male code specialists. Accordant with Evelina Fedorenko's view that the exact location of the language processing system varies marginally from brain to brain (while remaining in the left hemisphere), participants were first given language tasks to map their own systems. The same was done for a second candidate system, the bilateral multiple demand (MD) system, typically recruited for math, higher logic, problem-solving and difficult or novel tasks. Once this had been done, the conclusions were clear across Python and Scratch:

"We found that the MD system exhibited strong bilateral responses to code in both experiments, whereas the language system responded . . . weakly or not at all to code problems."

Shock. Results were near constant whether the problem being addressed involved math, string manipulation (i.e., words or letters), the output of a *for* loop or *if* statement—whatever. There was slight but discernible right-lateralization in the Scratch group, "perhaps reflecting the bias of the right hemisphere toward visuospatial processing." Otherwise, "Code processing is broadly distributed across the MD system rather than being localized to a particular region or to a small subset of regions," with "no consistent evidence of code-responsive regions outside the MD/language systems."

On the final morning we perform my scan, and when Anna Ivanova parses the results, they mesh with the rest, meaning "strong reliance on the MD network, little overlap with language." My brain appears to lean a little less on the left hemisphere than average, but not enough to raise eyebrows or be surprising given my different background and unorthodox path to coding.

In subsequent conversations with both Ivanova and Evelina Fedorenko, I try to get to the bottom of how the two studies generated such divergent results. Their theory is that the German team failed to acknowledge or account for the subtle differences in location of the language system within individual brains. These small variations led to misinterpretation of the data in Magdeburg, causing the MD and language systems to be confused with each other.

"It may be that their data is consistent with ours," Anna tells me, "but it was being interpreted differently."

Plausible.

But two features of this theory nag at me. The MIT study finds activation of the bilateral MD system in both hemispheres. So why hadn't something similarly bilateral been observed in Germany, even if it was misinterpreted? I could conceive of explanations for this: perhaps Janet Siegmund's team simply hadn't seen what they weren't looking for. Their methodology was different and may have been flawed. Yet none of this accounted for a discomfiting truth: as the one participant common to both studies, my scan results remained constant, with left and right hemispheres recruited to address the code snippets on each occasion.

Which begged the question, *What if Siegmund's team had not misinterpreted what they'd seen?* What if the telling difference had not been methodology, but the cohort under examination? I thought back to the German team's lead computer scientist, Professor Sven Apel, chair of software engineering at Saarland University, trying to find a delicate way of explaining his interest in the work.

"In our curriculum, we need to also educate students in English and German, because at the moment we don't do it," he told me. "We should work with them more on writing text and prose, like papers and presentations. Sometimes students . . . I don't know if you know this, Andrew, right, but some computer science students are hardly able to talk, to form whole sentences, when they come to us."

Up to now we've discussed language as being processed in the left hemisphere of most brains, but the truth is more complex. Syntax and structure—the sequential, logical, rule-based combinatorial stuff—happens there, but the facility for more advanced, creative uses of language, like metaphor, humor, thematic resonance, is broadly a product of the right hemisphere. We know this because people with impaired or inactive right hemispheres struggle (or fail completely) to understand these important aspects of discourse. Likewise prosody, the musical aspects of language such as tone and inflection, which (like music) is also generally right-lateralized. Asperger's syndrome is thought to involve impairment of the right hemisphere, but anyone with a very dominant left hemisphere may find communication with neurotypical minds difficult for all these stated reasons. Could it be that the German study subjects had been drawn from a pool inadvertently selected by computer science department recruiters for left hemispheric dominance, given the advantage this would confer in negotiating syntax and tight combinatorial logic in the early stages of a coding education? Could this also be what my friend had unknowingly described at Google?

When Jan notices me pacing the house muttering about cerebral hemispheres and abstraction, groping for some way to make sense of the patterns I am starting to think I see, she fixes me with an exasperated frown.

"You really do need to read *The Master and His Emissary* now," she says.

Jan is a voracious novel reader, so *The Master and His Emissary: The Divided Brain and the Making of the Western World* is not typical of her day-to-day reading. Or probably anyone's. *Emissary* is a five-hundred-page, tightly argued book of neuroscience and culture by an academic turned clinician named Dr. Iain McGilchrist. Despite its intimidating profile—and density—the book was a surprise bestseller on publication in 2009, with vocal fans including John Cleese of Monty Python and the erudite *His Dark Materials* novelist Philip Pullman, not to mention an army of reviewers, academic scientists and fellow clinicians. Jan is not alone in crediting McGilchrist with having revised the way she sees the world and has periodically championed his book as essential reading (just as I slip Richard Ford novels into her sock drawer when she's not looking). This is the first time she has used the word "need," though, which to my mind marks an escalation.

Her copy having been lent and never returned, I order a new one, and less than halfway through the introduction find myself rifling an email to the author, who now lives on the Scottish Hebridean Isle of Skye, suggesting we meet as a matter of urgency the next time we're both in London. To my relief, he agrees.

If the machines deign to upload only one human mind to their cosmic floppy drive come singularity time, they could do worse than Iain McGilchrist's. As a boy he won scholarships to the elite Winchester College and New College, Oxford, where he took a 1st in English. Elected a Fellow of All Souls in 1975 for the first of three times, he taught English while exploring philosophy and psychology, growing ever more absorbed by the relationship between mind and body, before realizing he would not find the truths he sought in the humanities alone and trading academia for medicine. He trained and then became a consultant psychiatrist and clinical director at one of the world's most storied psychiatric institutions, the Bethlem Royal Hospital & Maudsley NHS Trust in London, taking time out to work as a research fellow in neuroimaging at The Johns Hopkins Hospital in Baltimore. He is a fellow of the Royal College of Psychiatrists and the Royal Society of Arts. These days, at the turn of his eighth decade, much of his activity revolves around a website repository for his output called Channel McGilchrist. For anyone interested in the relationship between science and the humanities, his artful takedown of Steven Pinker in a reply to the latter's essay on the subject in the *Los Angeles Review of Books* is unmissable. Both are posted to Channel McGilchrist.

We convene for lunch at the National Gallery in Trafalgar Square on a crisp fall day, after a farcical interval trying to find each other with remote help from a third party: two men able to navigate Freud but not a floorplan. He is small and fantastically professorial, almost vicarish, with a neat beard, owlish specs and a soft, deep voice I imagine having evolved over the years to soothe patients. In conversation he wears his intellect lightly, sharing it when invited but otherwise listening to a degree that—unaccustomed as most of us are to such a disposition—can cause you to abruptly notice you're doing way more of the talking, when you came to hear *him*. At the same time he is playful, with a twinkle in the eye and enthusiasm easily tweaked by mention of something he is interested in, which includes almost everything. One of the most unexpected side effects of my coding studies, I tell him, bestowed by some of my Python buddies including Nicholas, has been an interest in Bach, who I used to think too mathematical. Just as my favorite revelation about the Apollo astronauts was that Neil Armstrong took an album of theremin music to the moon, so the most delightful detail from *The Master and His Emissary* to me is the possibility that Bach's contrapuntal music is arranged in such a way that our brains are compelled to hear it differently every time it plays. I mention this ("Yes, I really think it might be true!") and we're off. After half a lifetime lending himself to children needing mental health support; to sufferers from schizophrenia, psychosis, epilepsy and eating disorders; of directing the Acute Mental Health Services department and a hectic Community Mental Health Team in a diverse area of South London, you would expect McGilchrist to train a humane eye on the world. Which he does. But don't be fooled. Salving as his presence and elegant prose may be, there is an unflinching ardor to his work, the reasons for which become more apparent the further you delve into it.

There are several things I want from our conversation. McGilchrist is about to publish what he says will be his last big book, *The Matter with Things*, a work of empirically grounded philosophy built on the groundwork in *Emissary*, ten years in the making and as readable as a work running to 1,400 pages over two volumes is ever likely to be. So my first question, having lugged *The Matter with Things* across town on a bike, concerns a treatment plan for my back. But I also want to examine how his ideas might bear on the abstractions of code from a psychiatric, neuroscientific and epistemological point of view. Foremost in my mind is the question of whether what I may see in Google and Facebook and parts of the computer science establishment are local problems with local solutions, or whether something about the way we compute could be innately discordant with mind.

Why am I asking these questions of McGilchrist? Here's why:

The maturation of fMRI scanning in the twenty-first century opened the door to detailed mapping of the brain, but certain of its quirks had long been known and puzzled over. Chief among these oddities was its bipartite structure consisting of two hemispheres, each capable of sustaining consciousness on its own, joined by a mediating *corpus callosum*. And here the head-scratching starts, because the corpus callosum's role as mediator is to a counterintuitive degree inhibitory, aimed at *preventing* crosstalk between the hemispheres. Curiouser still, each hemisphere responds to and controls the side of the body opposite to itself. In animals with eyes on the sides of their head, the right hemisphere controls the left eye and vice versa. In animals with forward-facing eyes (like humans, apes, some crows), each hemisphere responds to and controls the opposing *field of vision* in *both* eyes.

There are obvious questions to ask of this arrangement. The brain's power stands in proportion to its ability to make connections, so why would the greater connective potential of a single unitary organ be shunned by evolution? Similarly odd is the fact, as expressed in an emphatic asymmetry, that the hemispheres are neither backups for each other nor straightforward custodians of different mental functions. While both appear to be involved in everything we do, loss of or damage to one will change the way we do these things—to the point where, in extreme cases, a person's personality can change completely. So the hemispheres work together but take pains to remain independent.

The trad neuroscientific response to this vexing arrangement was to dismiss half the problem by presuming the logical, narrowly rational left hemisphere to be running the show, with the right confined to vague, emotional, arty-crafty stuff. Appealing as this view was to mostly male brain scientists, it still posed the question, *Why bother?* If the right hemisphere was so fluffy and inconsequential, wouldn't nature have lost or repurposed it?

No one believes the right-side sidekick theory anymore. But what is the coherent alternative? Before *The Master and His Emissary* there really wasn't one.

McGilchrist proposed a novel way of addressing the mystery of the bipartite brain. Instead of asking, "What does each hemisphere do?" (answer: everything), he asked, "*How* does each do what it does?" And suddenly the inefficiencies and weirdnesses fell away. Part One of *Emissary* marshals a mountain of settled science to suggest that the critical difference between the hemispheres—the unique contribution they each make to survival—consists not of a specific roster of competencies, but a distinct way of seeing, a *type of attention* paid to the world. In Darwinian terms, the basic conscious requirements of any creature with a brain are to eat and avoid being eaten. Yet it's easy to see how these two imperatives—eat, don't be eaten—require different, even conflicting modes of attention. The bird searching for a worm needs tight focus restricted to its immediate vicinity, with any broadening of view imperiling the task. What about the hawk in the tree or cat in a bush, though? Catching a worm but being scarfed by the hawk won't do, and neither will outsmarting a cat at the expense of starvation. How to circle this square?

McGilchrist's leap of imagination was to suggest that in essence nature provides parallel brains working in tandem, with radically different horizons and relationships to their environment. "Attention is not just another 'cognitive function,'" he writes in *The Matter with Things*. "It is . . . the disposition adopted by one's consciousness towards the world." And as philosophers and physicists have long observed, our mode of attention affects not just how we see, but *what* we see. So where the left hemisphere is concerned with "narrow beam, highly focused attention," the right offers "broad, sustained vigilance." This lateralized aspect has been demonstrated in creatures ranging from slugs to bees, spiders and cuttlefish—the latter an intelligent cephalopod that attacks prey in the right visual field (left hemisphere) while scanning for predators using the left visual field (right hemisphere).

My bluebelly friends have skin in this game, too. McGilchrist describes an experiment in which lizards each had one eye patched and were placed in proximity to a simulated predator. As one would expect, subjects with their right eyes (left hemisphere) patched monitored the potential threat with their left eye (right hemisphere). But lizards with left eyes (right hemisphere) patched tried to use that eye anyway, even though it was covered. Every day I watch small birds detect jays in the right visual field (left hemisphere) and turn their heads to examine the backyard brutes with the left (right hemisphere). Crows have been found to lead with the right eye (analytic left hemisphere) in selecting a stick for use as a tool even where using the left eye would be easier. And the same has proven true of every animal with an identifiable brain yet studied. Moreover, this near-universal lateralization can be traced back six hundred million years to the evolution of the nematode *Caenorhabditis elegans*, which has a nervous system of only 302 neurons. Nature would not have smiled so unreservedly on this hemispheric scheme if it conferred no advantage.

Again, McGilchrist's key philosophical point is that the type of attention paid to the world determines "the nature of the world that is perceived." To a crow with a stick, everything looks like a grub, we might say, and the same is true for humans. Our left hemisphere controls the right hand, with which most of us grasp, along with "those aspects of language which enable us to say we've 'grasped' something—pinned it down, and help us manipulate, rather than understand, the world." For these reasons, the hemispheres have what might be thought of as distinct "personalities." McGilchrist continues

> *[The left hemisphere] sees little, but what it does see seems clear. It is confident, tends to be black and white in its judgments, and jumps to conclusions. Since it is serving the predator in us, it has to if it is to succeed. It sees a linear relationship between the doer and the done to, between arrow and target.*

The wide-open right hemisphere, by contrast

> *is designed to look out for all the rest—whatever else might be going on in the world while we are busy grasping. Its purpose is to help us* understand, *rather than manipulate the world: to see the whole and how we relate to it. It is more exploratory, less certain: it is more interested in making discriminations, in shades of meaning. Since it is serving the survival instinct and the social animal in us, it has to be if it is to succeed. All relationships in this hemisphere's world reverberate, changing both parties, and there is no*

> *simple linear cause and effect. Its attention, one might say, is not so much linear as in the round.*

We might think of the right hemisphere as concerned with context, relationships, direct connection with its environment—of which it feels itself part. Against this global outlook, the left hemisphere specializes in detached analysis, divorced from context in order to sharpen focus and better parse detail: it is—McGilchrist actually says this—the hemisphere of *abstraction*. These two dispositions taken together are indispensable to us. Under most circumstances, we would expect new sensory input to be fielded by the globally inclined right brain, passed to the left for analysis and returned to the right for recontextualization, without which its value is likely to be circumscribed. As McGilchrist puts it:

> *These two ways of seeing the world are each vital to our survival. We need to simplify and stand apart to manipulate things, to deal with the necessities of life, and to build the foundations of a civilization. But to live in it, we also need to belong to the world and to understand the complexity of what it is we are dealing with. This division of attention works to our advantage when we use both. However, it is a handicap—in fact, it is a catastrophe—when we use only one.*

The science McGilchrist recruits to make his case is well established and robust, drawn from multi-decade observations of patients with brain insults, trauma, stroke and tumors; from the more recent evidence of brain imaging; from neuropsychological experiments on normal subjects and post-corpus callosotomy, a treatment of last resort for severe epilepsy that involves deliberately severing the corpus callosum. All the same, his holistic interpretation of the science, challenging the atomistic traditional view of hemispheric differences as a collection of unrelated curiosities rather than "two distinct, entirely coherent versions of the world," has taken a while to bed in with a neuroscience community invested in established ways of seeing ("New ways of seeing always take time to be accepted," he reminds me over lunch). Some compare his insights to those of Darwin and Freud, while others remain dismissive. When I mention *Emissary* to Evelina Fedorenko, she tells me quite reasonably and with a smile that "I'm interested in things I can prove experimentally." More categorically, in an excellent 2019 documentary on the book called *The Divided Brain*, Dr. Michael Gazzaniga, Director of the SAGE Center for the Study of the Mind at the University of California, Santa Barbara, contends: "the brain is as mechanical as clockwork. A famous English physicist said that—let's just get over it"—a claim that even these few years later strikes me as impossible to stand up. Meanwhile, the equally garlanded Rutgers University professor Dr. Louis Sass tells the filmmakers, "The idea that there is a distinction between those two perspectives seems to me correct, and I see it all the time in my own field of clinical psychology."

Taken together these opposing views mirror the dichotomy *Emissary* describes, between a mechanistic Newtonian view of the universe and a more complex, dynamic, fluid—dare we say *Blakean*—one. It's clear to me, bolstered by the emergent support of big figures like V. S. Ramachandran, whom the evolutionary biologist Richard Dawkins has called "the Marco Polo of neuroscience," that the conceptual drift is toward the polymath McGilchrist.

The Matter with Things provides a slate of twenty differences between the two hemispheres, noting that philosophers including

Pascal, Spinoza, Kant, Goethe, Nietzsche, Schopenhauer, Scheler, and Bergson (who saw "two different orders of reality") had intuited them long before science. Many of these differences will be obvious by now, but number nineteen flips a switch in my head.

"The right hemisphere is better at seeing things as they are pre-conceptually," McGilchrist states,

> *fresh, unique, embodied, and as they "presence" to us, or first come into being for us. The left hemisphere, then, sees things as they are "re-presented," literally "present again" after the fact, as already familiar abstractions or signs. One could say that the left hemisphere is the hemisphere of theory, the right hemisphere that of experience; the left hemisphere that of the map, the right hemisphere that of the terrain.*

There are some fascinating outward manifestations of this bifurcation. Handed a baby, most people will hold it with head cradled to the left, in which position the left eye, controlled by the right hemisphere—seat of empathy—is most engaged. I try this on acquaintances and find only two exceptions: one whose left eye was damaged in childhood and another who is a dear friend and sensational person, but not the first I would turn to with a problem in need of compassion (and yet, if I needed a fix, or rescue, there would be no one better).

Probably from a mix of genes and experience most of us exist on a spectrum between the extremes of hemispheric control. But it doesn't take much right-brain action to guess that an overdominant left hemisphere comes with advantages in certain limited contexts while creating grave difficulties in others. Reams of academic study suggest psychopathy/sociopathy and antisocial behaviors, up to and including diagnosed antisocial behavior disorder, to be associated with a "hypofunctioning" (impaired) right hemisphere and/or "hyperfunctioning" (excessively active) left, even if environmental factors bear on how these imbalances express. And, of course, there is Asperger's, which is thought to involve severe right hemispheric impairment. Where imbalance favors the right hemisphere, consequences tend to be more tightly drawn: depression is strongly linked to a hyperfunctioning right or hypofunctioning left brain, as is stuttering. McGilchrist offers a pen portrait of the left hemisphere unbound, in which its vision of reality

> *is of a world composed of static, isolated, fragmentary elements that can be manipulated easily, are decontextualized, abstracted, detached, disembodied, mechanical, relatively uncomplicated by issues of beauty and morality (except in a consequentialist sense) and relatively untroubled by the complexities of empathy, emotion and human significance. They are put together, like brick on brick to build a wall, so as to reach conclusions that are taken to be unimpeachable. It is an inanimate universe—and a bureaucrat's dream. There is an excess of confidence and a lack of insight. This world is useful for purposes of manipulation, but is not a helpful guide to understanding the nature of what it encounters. Its use is local and for the short term.*

Faced with contradiction or anomaly, the left hemisphere will often dismiss or deny the dissonant information rather than seek to form a synthesis, because its preferred truths are binary and fixed: offered a story with episodes jumbled, it will reorder them according to similarity rather than narrative sense. V. S. Ramachandran refers to the right hemisphere as "the devil's advocate,"

McGilchrist points out, since "it acts as an 'anomaly detector,' on the lookout for what might be erroneously assumed by the left hemisphere to be familiar." The left hemisphere excels in "fine analytic sequencing" and complex linguistic syntax, with a larger vocabulary than its partner. It wants to narrow the world down to certainty, where the right orients to possibility.

When I came upon this description, my mind flew to Anna Wiener's *New Yorker* article on *Hacker News*, specifically to her description of the pitiless chatroom foment.

> *The site's now characteristic tone of performative erudition—hyperrational, dispassionate, contrarian, authoritative—often masks a deeper recklessness. Ill-advised citations proliferate; thought experiments abound; humane arguments are dismissed as emotional or irrational. Logic, applied narrowly, is used to justify broad moral positions. The most admired arguments are made with data, but the origins, veracity, and malleability of these data tend to be ancillary concerns.*

Then I think of the open-source wars; of Linus Torvalds and the productive but friendless Richard Stallman, recently #MeToo'd then forced out of MIT for the persistent treatment of women as objects. ("Stallman won't be the last," wrote Steven Levy in a *Wired* piece entitled "Richard Stallman and the Fall of the Clueless Nerd.") I think of Mark Zuckerberg's seemingly unshakable confidence and imperviousness to reflection or remorse, his evident inability to grasp or care about the impacts of his own actions on others—and bemusement when the others in question react badly to things he does or says.

Finally, I am dumbstruck to consider, as McGilchrist and I finish what is certainly the most fascinating lunch conversation

I've ever had, that the urgency of the scholarship he poured into *The Master and His Emissary* and *The Matter with Things* emerges from the brain's *plasticity*; its extraordinary capacity for adapting to the input it receives—for *rewiring* itself in common metaphoric parlance—which implies that the code a Zuckerberg creates to make the world a better place *for him* will impel the rest of us to adapt to be more like him. In this manner, just as individuals can inherit or evolve hyperactive left or hypoactive right hemispheres, says McGilchrist, so at times can whole cultures. He thinks we are in such a time.

"Yes. I think the way of thinking which is reductive and mechanistic is taking us over. We're behaving like people who have right hemisphere damage," he says. "It's made us enormously powerful, enabled us to become wealthy. But it also means we've lost the means to understand the world."

Is Google, are Facebook/Meta, Uber, Amazon and some computer science departments replicating their own imbalances at global scale through recruitment bias? There is no way to be sure right now. But we can ask what indicators we might see if they were.

We would expect to find far more men than women, especially among technical staff. A higher-than-average proportion of technical staff might display marked autistic traits, though that would not of itself be the problem. One would anticipate a corporate confidence bordering on recklessness and a cavalier attitude to breaking things—qualities shared with psychopathy and forms of the other dark tetrad subclinical personality traits. When insiders raise big-picture questions about the impacts of certain products or services, they and their ideas would likely be attacked and ejected rather than accommodated. Feeling boxed in, other big-picture thinkers would leave, further concentrating the institutional skew. You might be confused to find highly educated adults talking about other people, or pronouncing on society, as if they were children parsing the grown-up world by glib inference and assumption, having never lived in it themselves.

What else? You might note activities like clandestine plans to upload and make available for free all existing books, knowing this would destroy publishing and therefore *books*, the last repository for deep thinking. Met with outrage, a senior engineer might step forward to explain that the pages are being scanned for the benefit of machines rather than people, as offerings to the singularity. Told such action would be theft, a lawyer might remind anyone who needed reminding that his company's leadership "does not care terribly much about precedent or law." This enterprise would be identical in every respect to Google's covert Project Ocean.

Alternatively, a magazine reporter might wonder if a Big Code titan should consider directing its unprecedented computing power at curing cancer rather than a vague "AI" fantasy, only for a founder to muse:

> *One of the things I thought was amazing is that if you solve cancer, you'd add about three years to people's average life expectancy. We think solving cancer is this huge thing that will totally change the world. But when you really take a step back and look at it, yeah, there are many, many tragic cases of cancer, and it's very, very sad, but in the aggregate, it's not as big an advance as you might think.*[1]

1 Harry McCracken and Lev Grossman, "Can Google Solve Death?" *Time* magazine, September 30, 2013. https://time.com/574/google-vs-death/. The quote is from Google's co-founder and former CEO Larry Page.

Anyone who regards people as more than data points would be struck by the flaws in this logic. Not everyone dies of cancer, so three years added to *average* life expectancy translates to many more for those unlucky enough to be afflicted. Given that almost everyone loses somebody they love to a cancer, the sum of anguish attributable to it is vast, and "solving" it would be profound. Even the word "solve," drawn from the binary language of computing abstraction, would suggest Google cofounder Larry Page understood little of life's rich interconnectedness outside the company campus.

Or perhaps a CEO could be found lauding an abrasive culture built on "the obligation to dissent" as a way of encouraging "smart creatives," those "aberrant geniuses who can grate on other employees' nerves." Anyone who had spent time in real-world creative environments would recognize this as a narcissist's view of creativity—knowing the real deal to be driven in most cases by ardent pursuit of a truth, whether aesthetic, scientific or social, and rarely if ever by a temperamental need to dominate or dissent for its own sake. Confronted on surveillance and privacy, the same CEO, *Google's Eric Schmidt*, might be expected to say something like, "If you have something that you don't want anyone to know, maybe you shouldn't be doing it in the first place."

You would almost certainly encounter top engineers like James Damore decrying attempts to recruit women engineers, who on average will have minds configured differently from his (when the subject of diversifying Google's workforce is raised, one Black female Google employee told *Wired*, "You pretty much need to wait about 10 seconds before someone jumps in and says we're lowering the bar.") In insisting software engineers cannot afford to be "people oriented," and that tech companies must "de-emphasize empathy" to "better reason about the facts," Damore was doing what most of us do when an environment in which we are comfortable, which suits our strengths and minimizes our vulnerabilities, is threatening to change in ways we may find inhospitable: he attacked the apparent agents of change. An important fact of Damore's presence at Google is that he never applied to join. Recruiters coaxed him away from Harvard after noticing his performances in online coding puzzles. In a largely sympathetic *Guardian* interview, the engineer claims to have received "spectacular" work reviews at the tech giant. He also claims, while at Harvard, to have been diagnosed with what was then called "high-functioning autism."

Only after Damore was fired did some colleagues question whether he had ever really understood his job. The engineer Yonatan Zunger, who by coincidence left Google the week Damore did, contributed to the ensuing debate with a thoughtful *Medium* piece, in which he agreed that wrestling with computing logic is an unavoidable and important part of learning to code and structure programs in the early part of a software engineering career. Thereafter, the "real engineering" begins, in proportion to which the luxury of hiding behind cold rationality recedes.

"Essentially, engineering is all about cooperation, collaboration, and empathy for your colleagues and your customers," Zunger writes.

> *If someone told you that engineering was a field where you could get away with not dealing with people or feelings, then I'm very sorry to tell you that you have been lied to. . . . Anyone can learn to code. The truly hard parts about this job are knowing which code to write, building the clear plan of what has to be done in order to achieve which goal, and building the consensus required to make that happen.*

According to Jack Poulson, the atomistic viewpoint Damore embodies runs deep within Google. When I describe my experience on campus, he reminds me why he felt unable to stay: by turning any pursuit of information about what you were building into a disciplinary offense, management sought to erect impermeable barriers between departments. Just such a contractual provision was used to terminate Timnit Gebru's colleague Dr. Margaret Mitchell. Narrowly focused, socially isolated staff are likely to be easier to manage in the short term, but Poulson isn't the only one with fears for the long term. Googleplex designer Clive Wilkinson now expresses regret over his best-known work, branding its cossetry of staff "fundamentally unhealthy" and even "dangerous." Central to the architect's concern is an understanding that workers incentivized to stay on campus, and whose off-campus life is lived primarily among colleagues, "are not really engaging with the world in the way most people do."

Neither is the contraction of imaginative scope confined to the obvious suspects. In his book *After Steve: How Apple Became a Trillion-Dollar Company and Lost Its Soul*, *New York Times* reporter Tripp Mickle presents the departure of legendary designer Jony Ive from Apple as part of the "left-brain triumph" of technocratic CEO Tim Cook. A contributing factor to Ive's decision, Mickle reports, was the reshaping of the company's board in Cook's left-brain image.

"Mr. Ive was irate that a left-brained executive had supplanted one of the board's few right-brained leaders," Mickle writes of one appointment. "'He's another one of those accountants,' he complained to a colleague."

Mickle is using the terms "left-brained" and "right-brained" in a colloquial sense, having no information on the neurological makeup of the people in his story, but the fact remains: the most farseeing and creative player was pushed out.

What are the most bone-chilling things we could see in an environment where the right hemisphere's presencing influence is absent or suppressed? The wealth and power of Mark Zuckerberg and his early investor Peter Thiel. People who know these men speak of stunted individuals with limited emotional range and capacity to connect. Elon Musk reportedly suspects his ex-PayPal partner Thiel of being a sociopath (while Thiel considers Musk a fraud and braggart). Either way, the appeal to these men of a fallacious category of "pure reason," unmoored from emotion or empathy, is not hard to fathom: for them the unmooring looks to be a given, their only possible state. When Thiel parrots one of former British prime minister Margaret Thatcher's most misunderstood claims, oft-repeated but nearly always wrenched out of context, that "There is no such thing as society, there are only individual men and women," his position is comparable to someone with protan color blindness declaring, "There is no such thing as red or purple, only yellow and blue." Thiel has been open about his desire to explode the very idea of society, rationalizing his urge as a love of freedom and efficiency. But isn't he just using his tech money to recast the world in a manner that compels us to adapt to him rather than vice versa? View his championing of Facebook, and latterly Donald Trump, through this prism and recent history starts to make sense. These are not the people George Boole would have trusted his legacy to.

CHAPTER

17

Apologies to Richard Feynman

When I am working on a problem, I never think about beauty. I think only of how to solve the problem. But when I have finished, if the solution is not beautiful, I know it is wrong.

—Architect, inventor and systems theorist Buckminster Fuller

By the start of 2020 I've moved from OpenOakland to its sister Code for America brigade across the Bay in San Francisco. I loved OpenOakland, which was warm and welcoming and full of creative energy. But the only way to get there was via some of the most congested roads in America, swelled by rents that drove workers ever farther from the city. The group was attuned to this issue: most lived locally and cycled or walked to meetings. I could no longer justify the journey. In the event, my reluctant switch would prove fortuitous, supplying an unforeseen and improbable view of the positive force coders can be—and of the reserves of thought and skill required to do their jobs well.

The San Francisco brigade met Wednesdays at Code for America headquarters on Mission Street. I persuaded Sagar along after work one night and we were both impressed. Code for San

Francisco was bigger and more formal than its Oakland and San Jose satellites, with a deep pool of international talent, deft management, and an enticing portfolio of projects. Larger premises allowed members to mingle and eat together before getting to work in siderooms, while project reps presented ongoing work to newcomers, making the onboarding process quick and organic. In my dreams Sagar and I used that first night as inspiration to start a Marin County brigade, but I should have known he didn't have the bandwidth, that his passion projects tended to things like root-and-branch rebuilds of American democracy, through an app, written in Java—or else cool model trains. I knew he would help if I needed it down the line, but for the time being I was on my own.

I watched. Waited. Flirted with a couple of projects. Made friends and began to look forward to meetings. Until in March the world seemed to shudder and unspool like a snapped reel of film. From that point everything changed.

North America's first case of the virus turns up in the Bay Area. Handshakes and hugs commute to elbow bumps at meetings. Soon we're retreating to Slack and Zoom and having to relearn life. A new member dials in from a Mexican beach. "Wish we could be with you!" we gush. She laughs and shows us how to change our backgrounds, then we laugh as single beds-fridges-frayed-posters turn to sunsets, Death Stars, cats. Confusion abounds in this new virtual life. Over things like how to mute and unmute. When to speak. Basic stuff to worry about. A rare instance of continuity comes when JavaScript creator Brendan Eich angers staff at his new company with fallacious claims that masks don't work and the eminent director of the National Institute of Allergy and

Infectious Diseases, Dr. Anthony Fauci, is "a serial liar." Scientists say a vaccine for idiocy is still some way off.

Meetings grow and intensify as people look for ways to help with the pandemic. Civic coders turn out to be good in a crisis. Used to ceding egos to the needs of a group and facing problems without complaint. Maintaining sophisticated infrastructures for sharing. I'm not alone in feeling lucky we have each other right now.

A doctor drops in to explain that general practitioners are being left behind in the scramble for personal protective equipment (PPE). A team forms to explore ways to centralize the donation and distribution of PPE so it can get where needed. Ambulance drivers no longer know where to take noncritical patients as hospital beds fill with Covid sufferers. A team forms to design a simple phone app to identify capacity in advance. Then a front-end developer named Kengoy Yoshii brings a bigger idea. Coders in his native Japan have produced a Covid dashboard for greater Tokyo. He shares his screen and we see a clean graphical representation of the virus's progress through different districts. There are stats for cases. Hospitalizations. Deaths. Availability of services and local mandates. Kengoy hardly need explain why the Tokyo site is relevant. The Bay Area consists of nine counties with their own systems and statuses on the ground. Movement between the counties is perpetual. But there is nowhere to get a snapshot of conditions and trends across the whole region. Tokyo has provided a template.

After the main meeting we disperse into virtual "breakout rooms" for individual projects. For me the choice is easy. The new Covid dashboard team is abuzz. A well-liked organizing committee stalwart named Josh Freivogel will project manage. We introduce ourselves and declare our skills. A discussion of goals leads to the formation of three teams. Kengoy will be the technical lead and oversee a front-end Development team. This means coding and populating the website. He will be called "Kengo" within the group. Meanwhile Data Fetch will write the "scrapers" that pull data we need from other sources. I'm pleased to see a top Pythonista I know and admire on this team. Even he will be stretched at times. Data Fetch is where the real seat-of-the-pants programming goes down. By contrast, "UX" stands for "User eXperience" and concerns how the site feels to use. Without their work our code means nothing. I go away as breathless as everyone else. So this is what it feels like to be present at the start of a project.

The Slack channel hums as teams get up to speed. Thoughts and ideas fly. Logan from Data Fetch pulls a generic scraper from another open-source project to use as a guide. Watching the GitHub repos is fascinating. The scrapers turn out to be nightmarish. A couple of counties have APIs (application programming interfaces) to pull data from and these make things easier. The rest employ a menagerie of weird interfaces that require bespoke scrapers. For Data Fetch this is laborious. And never-ending. Whenever county developers make a change they break one of our scrapers. *Why are there no agreed standards?* We make plans to raise this with the state when the pandemic passes in a few months' time. Urgent calls for help become routine on the Data Fetch Slack channel. Counties also group data differently. What to do when one uses the age range 20–30 and another 25–35 and yet another 25–45? Before it didn't matter. A superbright female landscape architect turned coder comes up with some clever formulas to cope. Elaine was seconded from Code for San Jose, and I've been studying her work in our repo. She's *good*. But these things all cost time and the team is under strain. Faced with such intensity I remember the Google dissident Jack Poulson telling me that sometimes the best way to help is by not writing code. Truer in my case than his, I suspect. With a pang I decide to concentrate on areas I'm better qualified for until the scraper wars calm down.

Through the technical struggles it's easy to forget that our site is about information. The visuals are already winning. Nico from UX has refined the county graphs to hold reams of data yet be easy to read. And attractive. No small feat. A "splash page" sporting a solarized purple and orange Golden Gate Bridge is stunning. But I've been concerned about the quality of what everyone calls the "content." An FAQ section lacks focus and languishes unloved. Nicholas tells me he often shares this worry at the start of projects. We both think the word "content" cheapens the world. I raise the issue at the next meeting to find Josh way ahead of me. As part of his perpetual review of the project he perceives the FAQs to have bloated. They need to be sharpened and relocalized and monitored to ensure currency. He proposes the FAQs section as a fourth department. The considerable work of researching and writing and editing will naturally fall to UX but others will pitch in too.

I like spending time with UX. They're young and dynamic and good at their jobs. The questions they ask mix psychology with art and design. They use a platform called Figma to create "wireframes" showing how the site will look and behave. Front-end developers encode these wireframes into a working site. By contrast the FAQs live in a garden-variety Google Docs spreadsheet. We streamline the categories and rethink the questions, and after a couple of weeks we're on point. To localize the department further we consider a *Tales from the City*-style pandemic blog that could stand as a historical record. A little ruefully I agree to do the first one. (Just what I need! More words to write!) But before I can start work a more insistent need appears from out of far left field.

Everyone loves Kengo. So many things pass through him that he is always under pressure. Yet his openness and good humor are unfailing. I know this because (purely in the interest of science) I go to some lengths to test them. Early on I reached out to offer help but with little hope of use. Like most sites ours is heavy on JavaScript. Yet he now reaches back with a proposition. For the FAQs to be deployed they must be formatted in JSON. *JSON* stands for JavaScript Object Notation and is a lightweight data interchange language used to transfer and store information. In Pythonic terms a JSON object consists of dictionaries and lists of data nested within each other to whatever level a user requires. Incredibly this allows almost any real-world object or idea to be modeled. Its simplicity also makes JSON ideal for getting strings of text from a server to a browser. Kengo explains that at present he or one of his devs spends time encoding our FAQs into web-ready JSON. But as a member of the FAQ team who understands code maybe I'd like to do it?

I'm embarrassed by my pleasure at Kengo's request even if he has no notion of the chaotic force he is unleashing. Impostor syndrome is a thing in code. Even top-shelf coders claim it. Nicholas did once. *Christ*. But—loath as I am to pull rank here—I feel sure a case can be made for me being an *actual* impostor in this company. I hate this feeling and am ready to work hard to overcome it. So a first intimation I can be useful has meaning. I spend one evening getting up to speed on JSON. Another creating a neat web-ready JSON file for the FAQs. A third on testing and some sly re-editing of the content. Done. Now to GitHub to create a pull request. My first project "PR." A significant rite of passage.

Apologies in advance to Richard Feynman. But if you're a beginner and think you understand GitHub, you don't understand GitHub. I think I understand GitHub. Do the math. In our repo I open a PR for Kengo's attention. The next day—a Saturday—I print out our Slack exchange to keep as a souvenir because it runs to eight pages as we try to figure out where the blazes I *actually* sent my JSON. In the end it's found cowering under a rock somewhere. Begging not to be put through that again. Which is all I need to pledge once and for all that it won't.

Most coders will tell you GitHub is a little like JavaScript. Both are rooted in the minds of classic Stallman-issue Clueless Nerds™ and are ubiquitous products of the open-source evolutionary paradigm, by which we mean no one would *design* them to be as they are, but here they are anyway and everyone uses them because everyone uses them and they work. In the end familiarity breeds affection. I owe it to the pros I've somehow crash-landed amongst to learn to use our repo safely. A couple days' toil seems a small price to pay for neighborliness.

I find a fascinating *freeCodeCamp* article by Jacob Stopak called "Boost Your Programming Skills by Reading Git's Code." Linus Torvalds threw Git together in 2005 as a version control tool for devs working on his game-changing Linux open-source operating system, allowing the team to track and manage changes to the system. Stopak uses Torvalds's first interaction with Git, his first "commit," to lay bare how it works. The mechanics turn out to be simple. Git preserves a code file by saving or *registering* it to a Git file, creating a kind of snapshot. A programmer then registers changes at chosen intervals, first *staging* then *committing* them, at which point Git creates a new file called a *blob* containing the latest commits. The blob is compressed to take up minimal space, then used to generate a cryptographic *hash*, meaning a forty-character hexadecimal analog of the blob that will be used to uniquely identify it. (The term "hash" may be familiar from

Bitcoin, being an essential feature of the blockchain.) Git arranges past, present and future versions of the file into a "tree," creating a complete record of its evolution and allowing wrong turns or missteps to be rolled back to order. Imagine if we could do this in real life: existentialism would be out of business. Collaborators can now submit suggested new code without risk to the team. I get to the end of Stopak's piece and am shocked to feel buoyed rather than drained. I remember Wolf telling me there could come a point where I started to enjoy "the docs," but I hadn't known how to believe him. Something inside me has shifted.

Going into the files to see how the work gets done is empowering. Once there I see the simple truth that this is not sorcery. Like the blockchain, Git is a supple assemblage of established cryptographic ideas, an app built around instructions and data saved to files. That's all. For the first time I see the extent to which operating systems mimic the form of paper. Almost everything on a hard drive—including most problems—exist as data or instructions in metaphoric files somewhere. *Oh!*

As ever in code, one vulnerability exposes others. The best way to control Git is via the command line, so any blind spots there will tell. And now do. Which leads to similar exposures with the brutalist Bash command line language and the editor I've settled on, the modular, open-source VS Code. Time to get serious with these, too. How hard can it be?

Two weeks later I snap to as if from hypnosis, wondering where the time went on a task I thought would take two days. What have I done? Seems I've replaced Apple's default *Terminal* command line app with a third-party alternative called iTerm2 and, in a fit of optimism, decommissioned the venerable Bash command line language in favor of a newer rival called Zsh (spoken as "zee shell"). Which meant editing normally hidden Bash and Zsh system files in the darkest corners of my hard drive. Would I have done this if I'd known what it entailed? Never. But now that it's done, a joyous app called "Oh My Zsh" lets me customize and beautify my command line environment in hitherto unimagined ways. I spend an unforgiveable but ecstatic interval of time doing this—then again with my VS Code. One night Jan comes in, looks at my screen and gasps, "Wow, that's gorgeous." Now my spirit lifts every time I open my editor, and practice is a treat. Another day reading and rereading GitHub documentation, using its half-ass pervert's vocabulary of *branches*, *forks* and *pull requests* to bounce files between my editor and personal repo finds me starting to gain control. I feel like a moth emerging from its cocoon. Next time Kengo needs his JSON I will be ready to send it to the right place, or at least somewhere within crawling distance.

The work pays off. My next JSON file hits the target. I'm surprised at the way even menial responsibility fosters belonging. The mispronunciations and faux pas that betray a naif still plague my interactions with team members. Yet being at sea forces me to learn fast. Another group member is at about the same stage of development as me and I know she feels these operatic swings in confidence too. But just as I'm starting to feel settled a disturbing thought occurs: What if I go away or get absorbed in something else? What if the devs are still snowed under at the time? Who will manage the FAQ JSON? My anxiety is that any impediment to updating the information makes it less likely to happen and that stale information could have consequences during a pandemic. In retrospect the obvious conclusion takes a comically long time to appear. Here is a problem that can be coded away. That cries out for *automation*.

At first I'm not sure what to call a program that would pluck data from our FAQ spreadsheet and process it into a formatted JSON file. In fact it's a form of scraper. UX likes the idea of a program anyone on the team can run to order. So I mention the thought to our Data Fetch sage Rob, who could probably whip something up fast. Instead he thinks for a moment. Says:

"Yeah, that's a really good idea. But look, we've got our hands full right now . . ."

My head starts to spin as I intuit what's coming next.

"Why don't you do it?"

Shit. I order an O'Reilly tutorial book on writing scrapers with Python and arrange a chat with Nicholas, who mostly asks questions. Where is the data I want? How do I want to change it? Where is it to go afterward? The human answers are straightforward enough. I want to pull the text from our FAQ sheet and render it in a form Python can work with. I then want to reorganize selected text into a hierarchy of lists and dictionaries JSON recognizes, before delivering the result as a web-ready JSON file. Easy. To a person. But the computer is an alien with an alien's exotic needs. Nicholas and I discuss what these might be and yet I still spend the next week trying to understand the modules and methods my transformation will require. Then more days wrestling the pretzel logic that makes them necessary. By now I know these impasses usually hinge on an obscure piece of information I either don't have or haven't embraced. In this case the missing link is the behavior of *streams*.

Here's how the dance goes as I see it right now. A Python library called "requests" contains tools for requesting data from a url/web page. But we can't use the extracted data as-is because the

Python interpreter needs a Python file to engage with. Fine. Except that creating a file fathers a cascade of aggravation, forcing us to store it to our hard drive . . . therefore choose a location . . . reopen to extract data . . . decide what to do when we're finished with it. So instead of creating a file we keep our code ball in the air by treating the data as a stream. From this stream we extract the text and form it into a string that *behaves* like a file. A module called *StringIO* from the *io* ("input/output") library can do this. Google Docs spreadsheets are in a class of files known as *comma-separated values* (CSV) files and Python's built-in CSV library contains a toolkit of methods for interacting with them. The *csv.reader* method creates a "reader object" that can iterate over our *file-like object* and interpret it as a CSV document. Only after this series of switches and cutbacks are we positioned to write an algorithm to select and sort our FAQ text into formatted JSON. I see a *for* loop and bunch of *if* and *elif* (else if) statements. Relatively speaking the straightforward part.

Is it just me or is none of this fidget obvious? For the umpteenth time I'm struck by the degree to which a solution to a computing problem can appear logical and lunatic at the same time. The usual term for a program that acts as a tributary to a larger main program without being integral to it is *script*. I'll be trying to write my first script.

🐞 🐞 🐞

Deep breath. Sit down to work. Open editor. Install Requests library from the Cheese Shop. Doesn't work. Google to see if this has happened to others. Hours drift by. A brilliant article by a woman named Ray Johns starts with a ten-question quiz. Seven or more "yes" answers means you're not in control of your development environment. I thought I was by now but score seven anyway. *Argh!*

Johns's prescription is mastery of the *virtual environments* used to seal projects off from each other and prevent clashing. Python comes with a built-in virtual environment management module called *venv* but Johns sells me on a more sophisticated combo arrangement called *PyEnv-VirtualEnv*. At times over the next week I end up so far out of my comfort zone that I almost circle round and reenter it from the other side. But after a hairy setup PyEnv-VirtualEnv works like a charm. Installing it forces me yet deeper into my hard drive to find multiple rogue versions of Python and fragments of other programs. Requests hadn't worked because it was confused. Now it's ready to play.

I work through the script. It's called "The Jsonerator" until I'm advised to use the perhaps more mature *scraper_faq_sheet*. Internalizing the scraper's logic took far longer than anticipated. Now the formatting algorithm at its heart arrives with shocking speed. After a week of evenings I feel a rush of elation as the script runs and there it is: a JSON file that would have taken hours to create by hand in fractions of a second. *The script works!* But the most memorable part of the exercise is something I never dreamed of earning: a formal code review. As before I commit every sin against good practice and etiquette, feel each squirm of shame as if branded with an iron. But over three weeks' back and forth on GitHub Rob patiently and without a hint of judgment or irritation shows me the difference between a program that works and one that sings to a pro. I spend happy hair-tearing hours studying shortcuts and tricks and subtle extra features aimed at simplifying or adding flexibility. For the sake of clarity he suggests forming the data fetch mechanism into a function even though we're only using it once. I'm elated to find this also frees me from reliance on the StringIO module. Consideration for future readers is paramount throughout. Time-and-date-stamping each run of scraper_faq_sheet means figuring out Python's "datetime" module and a job-scheduling utility called Cron.

Setting up error messages for incorrect user input rounds off the formatting algorithm. Last comes the opening *shebang* line that tells the script where to find the Python interpreter it needs (and that looks like this: `#!/usr/bin/env python3`). The whole thing now runs to one hundred lines. More than a third comprise explanatory comments to help fellow developers and my returning future self.

Straining to resist pride during a pandemic I push scraper_faq_sheet to the project repo and go there to open a pull request. *Done!* Except that a warning bounces back. An intermediary platform called CircleCI monitors incoming code for adherence to project rules and has deemed mine uncompliant. My heart sinks when I recognize my most significant offense rests with a bullet I'd hoped to dodge. *Type annotation.*

Type annotation: to my eyes a means of nullifying Python's gracious ability to automatically recognize data types (strings, integers, Booleans, etc.) by forcing the types of all inputs and outputs to be specified in the function definition. Which aids clarity but is pug-ugly. With type annotation the first line of my function changes from:

```
def google_sheet_faq_data():
```

to:

```
def google_sheet_faq_data(sheet: str, gid:
str) -> Iterable[List[str]]:
```

Seriously? In the province of Guido's famously elegant language this is like having Aretha in the band and restricting her to kazoo. Irritated by the clutteration of code I secretly consider elfin and enchanting—and with Rob away backpacking—I Slack Logan to ask if this annotation malarky applies to me too. Turns out it does. So I spend another couple of nights reading the docs and experimenting. Which is how truth dawns that type annotation really *does* make sense for this project. Chastened and concerned Logan might not have enough else to think about I Slack him to announce my conversion. He says he wishes they used type annotation more often where he works. I thank him and open another PR. Rob returns and looks the new work over. "We're good to go ☺," he confirms and sweetly asks if I'd like to hit the "merge" button myself. Feeling as elated as I've ever felt, I do. I can code.

Moving from writing to code and back again has me rethinking the relationship between the two. The peculiar truth is that most of the time I don't enjoy writing, and I think most writers feel the same, even if this feeling is something we struggle to admit even to each other, mostly commuting it to a sardonic gallows irony. There are occasions, when you've managed to lose yourself completely and feel your ego melt away—the way an actor or athlete or musician does—that you brush a kind of nirvana, a sense that you scarcely exist. But part of what makes this transcendent state hard to reach is knowing it won't last and you'll be dumped like a sack of ferrets back into yourself before long. Worse, the abovementioned alternative routes to this state (acting, sports, music) are way more reliable. I saw the Norwegian novelist-memoirist Karl Ove Knausgaard explain this well when, asked by a New York book festival reader whether he finds writing hard, he replied, "No, writing is easy: getting to a place where you *can* write is hard." Only rarely does what I've written attain the glow of what I've seen in my head, or at least *think* I've seen but maybe never did, which appears like the clouds Turner painted over the Thames—as a kind of continuum, dynamic and complex—then hits the page static and linear.

The thing about writing is that it can always be better and it can always be worse, a complicated license to live with. A close friend who is also one of the best interviewers I know probably speaks for most of the trade when he quotes Dorothy Parker as saying, "I like *having written.*" We both feel embarrassed by how long it takes yet thrilled upon realizing when it's done—as for the first time, every time—that we now see the world a little differently. Sharing this sense is the joy of both our lives: we love it even as we hate it. But whether we do it or don't do it makes little difference to anyone outside our own families and circle of friends, who mostly experience it through a filter of irritation; and whether it's done well or badly exercises nobody but us, our editors and a few charming, wise reviewers.

In the beginning, the two pursuits—coding and writing—struck me as similar. In contrast to musicians, painters, mechanical engineers, no one who writes or codes would ever claim to have mastered their craft. One day I went out to collect the mail and bumped into Sagar electro-cycling up the hill on his way home from work. I stopped to say hi, but even before I spoke could see something wasn't right. Since moving in I'd learned that my neighbor had an extraordinarily open and generous nature and was nearly always smiling, but that these smiles contained as many gradations as the Inuit see in snow, and if this smile was snow it would be one of those grit-gray New York slushes only a plow finds joy in.

"What's up, Sagar?" I asked. "Is everything okay?"

A frown scrambled his brow.

"Oh, I just got chewed out by my boss," he said, going on to explain that a complex problem he'd taken on turned out to be even more complex than thought, meaning he was now behind schedule and stalling his team on an important new piece of software. That wasn't really the problem, though. The problem was that because he thought he could handle it himself, he charged in without ensuring his code would be easily comprehensible to others, meaning that now he needed help but couldn't get it. His boss accused him not of being a bad programmer—that would be absurd—but of failing to respect the code the way a sailor respects the sea, and thereby infecting his team *Rime of the Ancient Mariner*-like with hubris.

"This can happen even to someone like you?" I gasped.

"Oh yeah, it can. It's always ready to bite you in the ass. I should have known that, though."

He shook his head and went inside to carry on working through the problem. I could see it was going to be a long night. And in that moment I realized an important difference between my friend's work and mine—namely, that if I show a questionable piece of writing to a colleague or editor, they are likely to say "this doesn't work," but the judgment will be a metaphor. In Sagar's sphere, "doesn't work" means "Sagar, this *doesn't WORK*" and contains a heart-ramping presentiment of catastrophe: the written code is supposed to make something happen, usually automate a repetitive task, potentially for millions or billions of people, but in its broken state doesn't. So a hospital monitor clicks off; a plane falls out of the sky. And in this case there is no subjective debate about aesthetics, no possibility of saying, "Well that's just your opinion: my therapist's hairdresser thinks it's the best thing I've ever done." Not only is it useless and possibly dangerous, it may be compromising a larger program involving tens or hundreds or thousands of people. On the plus side, once a piece of software works, it works, at least for a while and in the right environment. Having seen how a complex program gets made, the idea of "committing" a piece of life-critical software to the actual world strikes me as more terrifying than it ever did. My respect for the people who do it well is boundless.

The deeper insight takes a long time to reach, though. It is that, while computer programs are written in linear form (and will continue to be until such time as we develop multicore brains), they *behave* more like the clouds of thought and impression I've always dreamt of handing readers but never will. Look at the way a web page lives in your browser, with words, images, fonts, dynamic JavaScript or even PHP scripts on far-flung servers all changing what you see—a mosaic of data tessellated from innumerable pieces found locally and on the web. Technically most things are happening in sequence, but given that trillions of instructions—the loops and recursions and asynchronous calls that confer dynamism—are being processed not at human speed but at light speed . . . well, to us the program might as well be that cloud, that continuum, forming and reforming like a murmuration of starlings, in one sense *alive* . . . and suddenly this impresses me as intensely beautiful, embodying that seraphic state of creative nirvana I've been dreaming of since I first picked up a pencil. The difference is that the program doing this probably won't have been written by one person—a few or many may have been involved. An unexpected and sobering thought: I will never be a good enough programmer to participate at this rarefied level, but if I were starting my life now, I might well choose computer code over prose.

CHAPTER

18

A Cloud Lifts

The way to solve the problem you see in life is to live in a way that will make what is problematic disappear.

—Ludwig Wittgenstein, *Culture and Value*

My preoccupation with code arose from a specific concern: that our evolved world was being reconstituted at dizzying speed in the image of the people recoding it, unelected adherents to a culture few of us understood—and which, like all cultures, embodied the bias of its progenitors. Worse, their trumpeted "disruptions" tended to be presented with Darwinian authority as inevitable. Were they?

Drawing closer I saw not one but three possibly overlapping engines of abrasion. In addition to the people who wrote an increasingly toxic global codebase, there was the broader system within which they worked. The third possibility ran a shiver up my spine, though. What if there was something about the binary structure of code that propagated itself in the minds of those who wrote and used it? Or that was simply more accessible to minds already skewed that way, affording them disproportionate power to shape a recoded world in their mental and emotional image—in effect replicating their mindsets like a virus? If our root problem was the logic of classical computing itself, then over time humans and their relationships could be expected to mirror the straight lines and silos undergirding it, leaving a binary patina on everything code touched. Here would be a problem on the scale of the climate crisis, with no straightforward solution, the only considered response to which would be terror. Start looking for signs of such a phenomenon and you'll think you see them everywhere.

There is no way to prove or disprove a direct relationship between the binary underpinnings of the microcosmos and the bifurcations we feel and see around us. My surprise at finding a significant causal relationship unlikely is exceeded only by my relief. Binary as the underlying substrate and lowest level of the stack may be, the levels at which humans engage are like mycelium fungus networks under a forest floor: rich, dynamic, complex—and likely to grow more so. While binary-tending minds may thrive on first contact, they have no innate advantage thereafter (even if by nature they might think they do.) If society seems to be drifting toward fractured, bipartite ways of seeing, it may be because Iain McGilchrist is right about a grand historical tug of war between the hemispheres—with code a contested part of the struggle like everything else—and/or because the demagogues who thrive in times of crisis find oppositional thinking useful to motivate followers and demonize designated foes.

I offered myself as a test case for the polarization I imagined, wondering if there would be a regimentation of my own mind in sympathy with code logic. fMRI scans are limited in what they can say about procedural changes within an individual brain, so the only tests for the changes I feared are subjective. What difference do I feel or do others see, if any?

With a jolt I realize four years have passed since my first "Hello World!" and that I've felt no constraint or regimentation. Rather, the weirdness of code and the discipline it demands; the acceptance that one always relies on the goodwill of others and is honor-bound to reflect their generosity back into the community—that vexation can be the seat of revelation, can be *exciting*... these approaches to the business of coding and living have been liberating to a degree I would never have guessed possible. At a minimum my mind feels more, not less, flexible. And by an order of magnitude sharper.

Am I imagining or rationalizing the qualitative difference I feel? I consult the one person in a position to know. Jan.

"I would say you have changed, for sure," she tells me when I ask, going on to cite several manifestations of what she means. One was when our oven broke and two separate teams of experts diagnosed two separate problems, touting solutions that didn't work. Tracing the logic of what each team said, I calculated that both were wrong: the problem could not be in the hardware; had to reside with software. Working through the program menu I found the problem and fixed it by changing settings. Another time our temperamental audio-visual system developed a pernicious audio fault at the juncture of hardware and software. I'd already stepped through possible quick fixes from the manual and knew there were none, so sat down one Saturday afternoon to sleuth documents and technical forums for clues as to the problem. Jan was in the kitchen an hour later when she heard the sound come on.

"My God," she told me later. "I've never seen you do that before. You just went very quiet and worked your way through the problem. That's definitely your coding: you've grown much better at thinking your way through things you haven't been trained for."

I had to agree. Could I have done these things before? Maybe with a gun to my head. *Maybe.* Would I have? Not a chance. Coders have taught me better ways to approach difficulty. To approach life. The subjective sense is of my mind having been brought into better balance, of more of myself engaging with everything I do. This is a powerful feeling, for which I am inexpressibly grateful.

The code for San Francisco Covid dashboard went live as the Bay Area PanDa (for PANdemic DAshboard). It was and for several years remained a sterling resource, an exemplar of Buckminster Fuller's view of functionality and beauty as inextricable. Looking back, I realize much of the code review for my script concerned aesthetics, whether Rob thought of it in those terms or not. By the end it was luminous in my eyes, type annotation notwithstanding.

By the time the PanDa hit its stride I was absorbed in something else. My first exposure to the audio engine and musical programming language SuperCollider ticked all the boxes for a platform I was primed to hate. It shared syntactic influences with JavaScript, to which it bore more than a passing resemblance, right down to a messy C-like insistence on semicolons to delineate statements—not to mention dozens of ways to perform a given task and myriad means of expressing the same thing. This was, in short, the Anti-Python, and facets of its logic screamed "No!" to me. Yet . . . within SuperCollider's obfuscatory folds lurked a mighty system whose offer I loved: SC3 was an unremarkable general purpose programming language until combined with the scsynth audio server, whence it invited you to make any kind of music you could find a way to imagine—while simultaneously coaxing you to imagine more. Because making beautiful music with ugly code seems contradictory, the language also challenged you to make the code beautiful in your own way, which meant that once written it was ineluctably yours. In the right hands it could read like a sonnet. I quickly found that to save my neck from the strain of raking between keyboard and monitor to police tangled syntax, my first task would be learning to touch-type. None of this would be easy. The clincher was, I didn't care. Here was the answer to the first question I was asked as a newbie, as everyone is: "What do you want to do with your code?" I expect to be doing this for the rest of my life.

Energized by the communities I've found and goodwill I've seen, I close the AND gate to my mind and allow optimism in. If the binary logic of code is not a decisive factor in the production of what we might call *psychocode*, code that emulates the behavior of a psychopath, then we're left with the two lesser but still serious barriers to advancement: (a) a demographic, cultural and cerebral imbalance in the engineering cohort, and (b) a nihilistic business incubus that outsources conscience and morality to The Market, theoretically justifying anything that turns a profit. Neither of these issues will be quick or easy to fix, but both look addressable through a combination of judicious regulation and education. The EU set the ball rolling with first attempts to establish ethical frameworks around privacy and data protection, disinformation, competition, and even algorithmic decision-making and "AI," setting an example regulators in the United States and elsewhere are following. OpenAI released their ChatGPT bot for experimentation at an early stage of development, opening an educative public debate in the process.

Focus has been on the Big Tech giants because they set expectations throughout the corporate stack. Most of the coders I know support clear regulation: the pernicious pushback comes from bosses and their lobbyists. Better, more constructive legislative incentives will advantage better, more rounded people likely to see the point of breaking tech's cycle of homogeneity and decadence. As I've discovered, these people are already abundantly present in the industry, they just need support. Our votes and consumer choices can help them.

The fight to bring more sustainable incentives to Big Tech, to enable long-term thinking and responsibility over short-termist opportunism—and to empower the thoughtful coder over the heedless—will be vicious and should be seen as part of a broader fight to guarantee a slate of rights currently under attack. There is a reason that at the end of 2022 Google, Twitter and Microsoft featured prominently in a roster of companies avowing support for democratic, sex and gender, reproductive and racial equality while providing financial support to politicians clamoring to undermine these things. If we're to continue as citizens rather than subjects, this is a fight we must win.

We should start that fight in schools. When I attended my first Bat Mitzvah, for the daughter of a Bay Area acquaintance, the only thing to strike me as odd about an otherwise beautiful ceremony was when the celebrant, in speaking of her sweet and laudable hopes for the future, lingered on the subject of "leadership." Afterward I asked around about the leadership thing, which didn't sound like the natural preoccupation of a thirteen-year-old, only to be told, "Oh, that's what they teach in school now." So, they teach all kids to be leaders? Who will they lead? What about those who don't want to lead or are not temperamentally inclined to it? From what I could see, "leadership" had devoured the notion of citizenship, aggrandizing the individual over the group. It took seconds to sense the fingerprints of mainstream Silicon Valley, of its tired "Great Man" view of progress and facile fetish of the "maverick" or "visionary (who made a gaming app)." This is one more heir to the disgraced California-forward pseudoscience of eugenics. After all, how does an Amazon CEO justify earning *6,000 times* the median salary at his company—against 351 across the economy (up from sixty-one in 1989 and twenty-one in 1965)—without recourse to the idea that he is special, superior?

Not content to incubate narcissism passively through their technology, Big Tech management's self-serving view of the world has been worked into the curriculum. Google calls this type of insinuation "social and intellectual capture," and there was a time when the self-conceit it betrays seemed charming. But as the *Silicon Valley* showrunner Alec Berg noted in explaining why he and partner Mike Judge pulled the plug on their popular show in 2019, "You get to this place where people are making very sustainable arguments that Facebook and Twitter and these other companies have torn the fabric of society irreparably. . . . And it ceases to just be a goofy, fun little show." Or as Judge told *The Hollywood Reporter*, Facebook's "move fast and break things" motto is "a little less cute now that they actually have moved fast and broken things."

Kids are on the front line in more ways than one. If code and its creators now comprise a Fifth Estate, then citizens need to understand their logic as surely as those of the legislative and executive branches of government, the judiciary and media. A grounding in algorithmic thinking illuminates the drives and imperatives manifested in code; the ways our behavior might be influenced by its unseen hand. School curricula are moving this way, which is to be applauded—with one grave caveat.

A recent flood of books, afterschool clubs, summer camps, and apps exhort parents to give their children a head start in the future work marketplace by force-feeding them code. Calls for kids to be taught programming from as young as three grow by the year. Which is understandable. But a giant mistake.

At this writing I know of no detailed research into the effect of early exposure to code on developing minds or the long-term health of the coding pool, but my own experience suggests it would be catastrophic for both. Young children have their hands full connecting to the everyday realm of the senses. Forcing them into the left cerebral hemisphere's abstracted, symbolic representation of that realm before sturdy relationships to people, places, things and emotions are established must risk stunting those relationships. To the objection that reading and writing are acts of abstraction, I offer an observation that most continental European educators introduce these skills later than is conventional in the Anglosphere, waiting until children are six or seven (notably in Finland and Estonia), with broadly better results. For children to whom socialization comes less easily, premature access to the tight sequential logic of computing could entrench difficulties by providing an unsustainable retreat from the complexities of people.

I am not the only one to feel this way. In a thoughtful piece published in *Slate* as "I'm a Developer. I Won't Teach My Kids to Code, and Neither Should You," Joe Morgan, the experienced developer of the title, explains why introducing kindergarten- or elementary-age children to code is likely to be counterproductive. Learning resources aimed at non-programming parents invariably present coding as "the new literacy," Morgan points out. Their kids will learn syntax, through exercises with binary right-or-wrong answers. But as I have learned, syntactic proficiency is the very least of a programmer's needs. And binary right-or-wrong solutions to computing problems are rare.

"Programming is messy," the developer reminds us.

> *[You] try something. See if it works. Try again. If a problem was straightforward, it would be automated or at least solved with some open-source code. All that's left is the difficult task of creating something unique. There are no books that teach you how to solve a problem no one has seen before. This is why I don't want my kids to learn*

syntax. I want them to learn to solve problems, to dive deep into an issue, to be creative.

What does he mean? Morgan recounts working to repair a wobbly chair with his small son, which began with a loose screw and ended in gleeful disassembly and reassembly. Another time they made sugar cookies, examining how the look, feel and taste of the dough changed with the addition of new ingredients, noting the qualitative effect of different combinations and how to maximize yield from a sheet of dough with a cutter. One of the crucial but ineffable senses a programmer needs is a feel for quality, he suggests.

That feeling of quality is the hardest thing for many developers to master. Well-designed code feels good to work with, and ugly code will make developers involuntarily cringe. The best developers learn to fuse abstract logic with the sensitivity of an artist. Learning to trust that aesthetic feeling is as much a part of development as any algorithm or coding pattern. . . . And every time you involve your kids when you work on something you value, you are teaching them how to do things well. You are preparing them to write code.

Nothing about the great coders I've met shakes my view that the very best thing we can teach children—regardless of what they grow up to do—is curiosity. Dijkstra considered mastery of a native tongue foundational for programming, too, while emotional intelligence is hard to gainsay as a necessity to working in teams. Play, art, aesthetic sense, empathy, ethics, interest in other people: these are deeply relevant to code and coders but seldom if ever addressed when programming is taught. Future research may prove

Joe Morgan and me wrong about early exposure to code. Great. But until then we might consider following the example of many Silicon Valley tech parents and treat code, screens and technology in general with the utmost caution when it comes to the young. That's if we don't want to foster the production of yet more psychocode. We might also note that in February 2022, DeepMind announced that their new "AlphaCode" algorithm had beaten almost half of five thousand engineers in a series of elite coding competitions. It would take a bold soul to bet that automated programming stops there.

Alan Kay, jazz guitarist turned creator of object-oriented programming in the seminal Smalltalk language, one of the most brilliant and humane minds to grace computing (a kind of Kurt Vonnegut of code), echoes Morgan's sentiments when he declares that "perspective is worth eighty IQ points." He maintains that CS undergrads should be learning math, science, systems, engineering, art and history as essential aids to good programming—and that programming should not just be geared toward future professionals.

"Our first English class in college, we're not aiming that class at people who are going to become professional writers when they graduate four years later," he mused on receiving a coveted Turing Award in 2003. "We actually think of the impact of the printing press and the new rhetorics and new ways of arguing that came with the printing press as something that is larger than becoming a professional reader or writer. I think the same thing is true of computers. Fifty years from now, this will not be controversial. Right now [computer science is] thought of, even at MIT and Stanford, with its great endowment, as basically vocational training with Java."

More than anything, Kay thinks we should be teaching "the romance [of] this unbelievably beautiful new art form."

A cloud lifts. Everything that needs to be done can be done. Properly understood, code and the code culture militate against a binary black-and-white view of the world. Both started from a place of expansiveness before retracting to defensive homogeneity like a sea anemone poked with a stick—a problem that can be fixed with sufficient will. A will may be emerging.

"I don't think enough businesses have come to the realization yet that we really do need to hire people who are good people," the Python Steering Council Member and Principal Software Engineering Manager at Microsoft, Brett Cannon, tells me. "More are catching on, though: I used to hear a lot about that '10X' or '100X' programmer who could have no communication skills and be the biggest jerk in the world—but, hey, who cares, 'cause they're productive! But it does seem like more companies are realizing, 'No, actually, teamwork is important,' which typically leads to those people also being able to communicate externally, because they just know how to communicate. Which suggests that there's hope, eventually, of getting there."

Cannon notes an "asymmetry" in the fact that open-source developers like him have implicitly bought into a social contract not necessarily recognized by incomers. But a contract is only a contract if all parties are signed up. Open source needs one, perhaps in the form of a basic training certificate in collaborative communication. Pretending dark tetrad personality traits don't exist and that everyone is of equal maturity favors only the trolls, whose magic cloak of anonymity must be removed. In 2021, NASA announced that its next generation of Mars Rovers would run open-source software—having previously used it to power a first helicopter drone flight on the red planet—which shows how important this inestimable creative engine has become. The talented people who make it work need protection.

Once again, aren't these teething problems in a new realm of experience? Psychologists say humans are springloaded to linger on negatives because focus on danger keeps us safe. So a whistleblower's July 2022 revelation that Uber's criminality and cant exceeded even the direst imaginings of critics stays at the front of my mind, as does an account the same day of how "AI" pricing algorithms are learning to collude, another warning that TikTok algorithms bombard young men with misogynistic posts, and news of the Chinese government using its equivalent of Microsoft 365 to censor books before they've even left their authors' laptops. Contemporaneous reports of an ambitious French plan to democratize large language model "AI"; of a virtual reality program called "Isness-D" that simulates psychedelic transcendence in the brain, opening a way to chemical-free treatments for maladies including obsessive-compulsive disorder, post-traumatic stress disorder, depression, and addiction; of DeepMind's AlphaFold algorithm completing its mission to predict the structure of every protein known to science, a phenomenal achievement . . . these, being positive, are saluted then forgotten. Technologies that work and are virtuous demand no sustained attention.

Have I been too pessimistic, then? In the end this is all about tendencies: As more code flows into the world, will our lives tend to get better or worse overall? I try to reopen myself to my environment, to see the everyday with eyes rinsed of doom scroll, hyperbole, science fiction. At the dentist I watch an X-ray being shared with the lab that made a problematic crown, can't believe my eyes as a technician diagnoses the issue and remotely guides my dentist through a tricky fix in real time. The magic of code, connecting people in useful ways. On the drive home a dashboard light warns of low pressure in a tire, reminding me that cars used to break all the time but hardly ever do now, thanks to code that monitors and regulates them; helps manage design, engineering and manufacture. At the DevCon conference in Oakland I meet a rep from an elite remote coder recruitment agency who speaks of an uptick in applicants from impoverished East African countries like Ethiopia and especially, amazingly, Djibouti, where pirated O'Reilly books are weighing against the lottery of geopolitics. Back at home, an email describes a machine learning–propelled medical revolution in diagnosis, treatment, and drug design. The latter sounds now familiar warnings about bias in ML training datasets and the technology's uneven distribution. But we are talking about these issues, which implies they will be handled—doesn't it?

Of course it does. Scientists have got our backs. Legislators are rousing, consumers awakening. Bipartisan consensus has formed over the damaging impact of social media on children: the conservative political analyst and academic Yuval Levin has opened a discussion on what to do about it. And I've hardly mentioned the Big Tech companies who recognize their power and try to use it well, like Slack and Salesforce. Give coders good things to code and they will code good things. Liberated from the *Hacker News* and haters, their culture of sharing has the potential to save us.

For the first time in eons I feel unconditional lightness. Until . . .

CHAPTER

19

Strange Loops and Abstractions

The Devil in the Stack

The answers are always inside the problem, not outside.

—Marshall McLuhan

I'm coding. Writing. Coding. Writing. For no specific reason I'm reading a book about the nature of consciousness, called *Gödel, Escher, Bach: An Eternal Golden Braid*, by Douglas Hofstadter. You can tell Hofstadter wrote *Gödel, Escher, Bach* when he was young because it's both fearless and original, using imagined dialogues in the style of Lewis Carroll to illustrate a long sequence of exhilarating ideas. At the heart of his take on the mystery of self-awareness, our sense of having an "I," is the bewitching notion of complex recursive feedback loops he calls "strange loops." At least one computing conference is named for these "strange loops," though Hofstadter claims to have no interest in computers as such and has expressed skepticism toward current "AI" models.

Despite its length and density, *Gödel, Escher, Bach* became an unlikely bestseller when published in 1979, winning a Pulitzer and a National Book Award and acquiring a cult status it retains to this day. Cut from the same cloth as Iain McGilchrist, Hofstadter went on to have an academic career taking in the study of computer science, cognitive science, artificial intelligence, psychology, comparative literature, linguistics, history, philosophy and the philosophy of science. His art has been exhibited and his piano compositions performed: he has translated works of literature into English from French, Italian and Russian (from the latter, the poet Pushkin). Rarely is so erudite a writer so playful. To illustrate the shifty concept of recursion he not only recruits Bach's canons and the work of Dutch artist M. C. Escher, but formulates the incontestable Hofstadter's Law, that "It always takes longer than you expect, even when you take into account Hofstadter's Law." Think about this statement for a minute and you see how funny and clever it is, how elegantly it illustrates a difficult concept.

And yet the thought that stays with me from *Gödel, Escher, Bach* is one I didn't expect. In the introduction to the twentieth anniversary edition, Hofstadter describes the steps by which his unorthodox first work took shape.

> *One of the key qualities that made me so believe in what I was doing is that this was a book in which form was being given equal billing with content—and that was no accident, because* GEB *is in large part about how content is inseparable from form, how semantics is of a piece with syntax, how inextricable pattern and matter are from each other.*

Content is inseparable from form. I spend a lot of time considering whether Hofstadter's statement has precisely the same meaning as the Canadian philosopher Marshall McLuhan's famous sixties adage, "The medium is the message." Eventually I decide they are not the same, but do describe a shared truth about the world. For McLuhan, "medium" means *tool* and tools are extensions of ourselves. The book is a medium, but so is the train, the cart, television, money and the internet. In saying that these mediums are their own message, that their form *is* their primary content, he means that the specific information conveyed by a medium like the internet comes a distant second in importance to the channels it opens, behaviors it enables, relationships it makes possible—the way it changes our being in the world; changes *us*. Put another way, mediums propagate their essences, their *forms*, in all who use or encounter them. This is what John M. Culkin meant when he elucidated his friend McLuhan's ideas by saying, "We shape our tools, and thereafter our tools shape us."

The outlook described by Hofstadter and McLuhan is less exotic than it might seem. The injunction to "Be the change you want to see in the world," whether attributed to Mohandas Gandhi or Arleen Lorrance, comes from the same conceptual place. So did the late great Parliament-Funkadelic showman George Clinton when, toward the end of his career, he inverted the hippy mantra (and title of his group Funkadelic's second album), "Free your mind and your ass will follow," to "Free your ass and your mind will follow." He was right. I always used to wonder why explicitly political rock bands or playwrights or filmmakers tended to disappoint their fans in the end, having failed to change much by the time fashion moved on, while—to use music as the example—bebop jazz or acid house (or Bach, for Hofstadter), which were radical in form before content, never did; seemed to alter their audiences' ways of thinking and being, their understanding of themselves, for life. Their most important messages were delivered through the fact and form of their existence.

When the embedded software mensch Jack Ganssle tells me ethical outlooks are inescapably embedded in the structure of programs—as my own code review at Code for San Francisco made plain—or when Sagar's colleagues at Salesforce remind each other that "A company exports its organizational structure," they are acknowledging the relationship McLuhan and Hofstadter articulate. We've seen how this can work with code in the examples of PowerPoint and digital music, but the motion McLuhan describes can be felt at greater scale, too. Most Britons now recognize their country's colonization by Russian oligarchs, its openness to money laundering in the decades prior to the Russo–Ukrainian War, to have damaged the polity as a whole. McLuhan would not have been surprised. The oligarchs' medium was crooked money and its message was the message of its distributor, Vladimir Putin: "There is only power and self-interest." The money's quintessence, *corruption*, replicated in everything it touched and may take a generation to cleanse. All perfectly predictable.

Which brings me to a question I haven't asked before because the answer seemed obvious. *What is the medium of computing?* What is the irreducible concept or conveyance without which it cannot exist? Prior to entering the microcosmos I would have defaulted to most people's assumption that the medium of computing is silicon. But we now know that in principle computers can be constructed of any material with two or more possible states—of steel (per Babbage), vacuum tubes (Tommy Flowers), marbles (Leibniz), chemicals (Lee Cronin); of sticks, bottle caps, dead crabs, lounging oligarchs. The medium of computing is not silicon.

Binary code was the next obvious candidate, which—*aha!*—made seductive sense in a polarized, binary-inclining world. My subsequent inability to credit a decisive link between the two (real though it may be in limited ways) left me comforted but confused. It shouldn't have. The new quantum computing technology eschews classical bits for "qubits" that harness the quantum mechanical property of "superposition" to represent zero or one or *both*: indeed, a computer founded on similar "ternary" logic was built in the Soviet Union as long ago as 1958. Only now do I recall that ENIAC and many sophisticated early machines used *decimal* logic. Would these machines have propagated their decimal structures? Hard to believe . . .

With a jolt I understand that I wasn't persuaded by a direct symbiosis between binary code and a polarizing world *because binary code is not the medium of computing.*

But if binary code isn't, what is?

The answer strikes like lightning. Abstraction. The medium of computing as presently understood, the irreducible conveyance it can't do without, is abstraction. Which means *the message of code* is abstraction. Bubbling into the world like water from a flooded drain.

And in that flash everything starts to make sense.

In "Counterparts," the ninth story from Joyce's collection *Dubliners*, the great author writes of his protagonist, "His imagination had so abstracted him that his name was called twice before he answered." Before computing, "abstract" was used in various ways to suggest something that had been removed from its natural place or detached from the thing it represented. This could involve being painted, excerpted, stolen or obscured. The computing definition of "abstraction" draws from these earlier meanings but is highly specific to itself. As we already know, it describes the act of representing a complex low-level system in a way that's easier to understand and work with, burying complexity under a new layer of code. Files are abstractions for bytes on a disc, allowing us to manipulate those bytes without having to engage them directly. In SuperCollider the "UGens" used to generate sounds are abstractions for mathematical calculations with signals on the server. Computerists half joke that "Every problem in computer science can be solved by another layer of abstraction." These layers of abstraction comprise the stack.

So integral is abstraction to computing, so taken for granted, that when I try to work out exactly how my Python gets to the metal—the bottom of the stack—it takes days of research to trace the steps. To greatly simplify, when my program runs, the Python compiler uses a process called *lexical analysis* to sift my source code and identify its building blocks (functions, lists, dictionaries, variables and so on). These are turned into *tokens* contained in a data structure representing my code, called an *abstract syntax tree* (AST). The AST is analyzed, optimized and eventually turned into a series of instructions called *bytecode*. This bytecode contains instructions for a simulated computer called the *Python Virtual Machine*, which is (essentially) what we mean when we speak of "the interpreter." In the standard CPython implementation of the language, the interpreter is coded in C and works by calling a range of prewritten C functions that instruct the operating system (Windows, Android, etc. . . .) to make things happen. Such C code is compiled to machine code, consisting of nothing but zeros and ones, and is sent to the CPU to be *executed*. Most users will be at most dimly aware of this byzantine procedure, which carries on above the level of Python, too, in the form of libraries, frameworks, dependencies, and so on. By contrast, in C you write your code and the compiler follows a similar procedure to the Python compiler but outputs machine code. No wonder C is faster and harder to work with: there's much less stack. If it were pancakes you'd send it back.

Abstraction in computing has certain intrinsic properties. It stretches the conceptual distance between source and signal, input and output, concealing chains of connection and causality. We see what goes in and comes out, not what (or how much) happens in between. One of its signature qualities is therefore to make the end points in any sequence of events *seem* close together while actually pushing them apart. Inside the microcosmos, in the province of light speed and symbolic representation, this is helpful, wondrous even. What does it look like in the slower, evolved social cosmos of humans, though—the cosmos of earth's space and time?

Ask the question and answers appear with the force of birds thudding into glass. Globalization, with its ever more tortuous and brittle supply chains. The arabesque mechanisms of corporate finance and plutocratic tax avoidance. The disinformation industry, with its baroque stylings and conspiracy theories designed to stuff unseen—but *felt*—economic and social black boxes with narrative rationale. Social media companies that tear the notions of friendship and "connection" to fraying threads, dissociating relationships from physical space and ceding agency to yet more covert boxes whose precise agenda in a given instance is unknowable. The stage managed anarchy of algorithmic stock markets, where most shares are held for fractions of a second and participants are shrouded in mystery; where ordinary investors are cast as patsies and any connection to capitalism's original promise of directing surplus wealth to where it can be productive has long since been abstracted away.

These platforms work by propagating black boxes that distance source from signal, input from output to the point where the source itself is indistinct. By presenting segregation as engagement, abstraction feeds a dangerous emaciation of empathy. All these facts of twenty-first-century life are mediated and made possible by computer code—and mimic its form, fracturing civilization while appearing to gather it closer. Code is obviously not the only conveyor of abstraction in the world, but as an accelerant it would be hard to beat.

Now we start to see more generalized cumulative truths about our abstracted selves. Just as everything on a hard drive is effectively equidistant (given that electricity travels at nearly the speed of light), so everything in a highly abstracted human environment starts to feel present to the point of claustrophobia—yet still somehow ambiguous, intangible, occult, as if manipulated by unseen forces. Which it is, because that's what abstraction *does*. In the human realm, an older word for this kind of abstraction is "alienation." Aldous Huxley's prophetic *Brave New World*, Charlie Chaplin's symphonic masterpiece *Modern Times*, the first English translation of Franz Kafka's *The Trial* . . . all were against industrial, bureaucratic and intellectual alienation, all appearing within a few years of each other at another time of swelling abstraction in the early to mid-1930s, as the first computers were emerging to wage a war these works portended.

This is not what I expected to find when I set out to study code. It takes a while to get over my shock and contemplate what to do with this information. Precedents are less than promising. Had modest adaptations been made when scientists began to flag a dangerous link between carbon emissions and global heating from the late 1960s on, the world would be a safer, saner place right now. Can we do better with code? What would it take?

Professor Neil Johnson, the theoretical physicist last found floating a new science of worst possible outcomes, led research claiming mathematical proof that code fixes Facebook proposed for its algorithms would inevitably worsen their effects. We now see a deeper sense in which this may be true. Wherever a real-world problem in a large open system is perceived to be one of alienation or abstraction, we should expect any attempt at a coded remedy to instead act as a multiplier, amplifying the effects we seek to abate. That's even before the loops and recursions go to work on concentrating them.

We see something important here. Closed software systems—like the one my dentist used; like DeepMind built to control a fusion reaction; such as NASA will employ on their next generation of Mars Rovers—all accord with what computerists like Guido van Rossum and Andy Herzfeld expected when they set out to compute. They avoid the unpredictable network effects, the strange loops and chaotic resonations, observed in large open systems. Discrete algorithmic programs can still do harm, of course, most obviously when used to automate decision-making in relation to individuals. Algorithms designed to sift job or college applications, loan approvals, the likelihood of committing a crime and therefore suitability to being searched or arrested . . . these have been shown to exacerbate existing iniquities and should be banned until proven safe. We humans might be prejudiced or lazy or stupid sometimes, but unlike algorithms we can be compelled to explain ourselves. Recommendation engines need to come with the equivalent of nutritional labeling on food (something the food industry fought tooth and nail, lest we forget). How are their recommendations arrived at? Are any paid for? What data is being fed to them? We need to understand that over time the logic of recommendation engines is to narrow individual and collective scope.

We also need to stop talking about "AI." It's machine learning. The process by which the machines "learn" is important and needs to be acknowledged as such. A nitpicker might note that the verb "learn" is algopomorphic in this context, suggesting a more human-like ability than is the case. We've seen machine learning programs do precious things within closed systems (and also worrisome things, as per high-frequency trading). But as I write we don't know enough about their effects within large open systems to let them colonize the space between us. Machine learning algorithms that interact directly with people should therefore require a license, to be granted only when applicants are able to prove their work safe in the intended environment, much as happens with new cars or planes or drugs. Predictive algorithms such as those creeping into law enforcement armories run counter to the rule of law and have been shown to be racist: they have no place in a civilized society and should be forbidden.

The vast, highly abstracted open systems comprising social media present the greatest challenge. A checklist of minimum necessary reforms is easy to assemble. The whistleblower Frances Haugen's eloquent US Senate testimony made plain that Facebook/Meta had been grooming kids for monetization and knew it was harming teenage girls (particularly via Instagram). We are abdicating our responsibilities—and will pay a price down the line—if we allow this to continue. Children should not be allowed inside present social networks any more than they are allowed to drink, smoke, drive or marry. If someone designs a kid-safe variant, let them apply for a license to run it, with the burden of proof on them, as for a pharmaceutical product. Parents, teachers and child psychologists need to be in the loop.

The laws applied to online portals must be harmonized with those that apply offline. Social media companies are media companies and should be subject to the same legal frameworks as newspapers and magazines, which means taking responsibility for what they publish. If they can't do this, they should leave it to professionals who can. Legal consistency won't stop disinformation but will help hold its purveyors to account. For the same reason, anonymity should no longer be allowed within large open systems. Twitter has often been referred to as a digital "village square," but the point about village squares was that you could see and weigh the trustworthiness of those you interacted with, could look them in the eye.

We need to get serious about moderation, too. A millions-strong community like *Hacker News* being moderated by two beleaguered people is absurd. Early social networks, in the form of bulletin board systems (BBSs), MUDs and MOOs, were closed systems containing discrete communities that set their own rules and dealt with their own problems. Even these could be a handful to manage, but they came with no danger of toxic overspill or contagion and had no commercial dimension to exploit. Most of the work I see coders doing on next-generation social networks resembles these first iterations. We know present networks are designed to be addictive: a lecturer named Nir Eyal taught a course on "behavioral engineering" at Stanford and wrote an insidious book called *Hooked: How to Build Habit-Forming Products*. If Facebook can be designed to "exploit a vulnerability in human psychology," then laws may be framed to identify and discourage the mechanisms of such exploitation (ideally by advantaging good practice over bad). The point is not to stop people from seeing things they want to see, it is to help them see what *they* want to see, not what algorithms calculate most lucrative to their masters. A useful rule of thumb for us all is to question and closely examine anything purporting to offer convenience as its main selling point. Who offers the convenience and why? What's in it for them?

But a shopping list of reforms is not the point. The point is to open a broad discussion about what we want code to do for us and what we don't. In the end, this will be a discussion about abstraction. We can start by assuming *every problem in society will be heightened by another layer of abstraction*. Big Tech pushback and lobbying against moderation of their power will be as intense as the motor and oil industries' decades-long war on climate science, and for the same reasons. Good technology has nothing to fear and, like us, everything to gain. In a pamphlet rousing the American colonists to rebellion against the British in 1776, Father of the Revolution Thomas Paine excoriated fainthearts and doubters with the words, "We have it in our power to begin the world over again . . . how trifling, how ridiculous, do the little paltry cavilings of a few weak or interested men appear, when weighed against the business of a world." Unlike the colonists in 1776, we have drifted beyond the luxury of choice: we *have to* begin the world over again. And soon.

Acknowledgments

Acknowledgments always strike me as the strangest part of a book. By their nature always written before a work has entered the world, it's easy to see how they could read—to anyone lucky enough never to have tried to write a book—like an Oscar acceptance speech without the audience or Oscar. Most of this is about relief. You were writing a book; now you're not. Anyone who helped change that particular zero to a one is going to be close to your heart.

So here is a list of names that, taken as a whole, probably means little to anyone other than me, and yet my gratitude for—and pleasure in—their generosity is the truest thing in these pages. When all is said and done, the experiences we shared along the way are what I'll remember.

I want to start with a celebration of the journalists, authors, bloggers, podcasters, documentarians and academics who cover the areas lassoed into this book. In most cases their very existence is a rebuke to the blanched and diminishing idea that people can only be motivated by money or advantage, and their curiosity in and of itself makes the world a better place ("Curiosity requires a kind of gentleness," as someone puts it herein). Science and technology have never been better written about. We all benefit from this and are well advised to support it at a time when deep thought and nuanced discussion are under attack from all sides. The online "Notes & Sources" section of this book stands as tribute to the brave chroniclers and investigators. Gods bless you all. Never give up.

Specific to this book were the coding groups I encountered and sometimes entered. Within Python I found an extraordinarily warm and vibrant community, full of remarkable people, most trying to do their best by the world. PyCon is now one of my happy places, where I go not to work but to see people I love, something I would never have imagined possible. Similar can be said of the SuperCollider, SonicPi, algorave and Code for America clans. Special thanks must go to Naomi Ceder and Russell Keith-Magee for reading parts of the manuscript for accuracy (though any failings in that regard remain most assuredly mine), and to Rob Brackett of Code for San Francisco, from whom I learned so much, and whose patience—in common with others mentioned here—lights the world like a beacon. When I was worried about the "binarification" of my mind via code, I had only to look at these people for an inkling that it might be alright in the end.

Many of the above are also educators. No one reading this book will need persuading of *their* importance. Quincy Larson of the amazing freeCodeCamp, Nick Bergson-Shilcock of the Recurse Center, and the gifted communicators of code Eric Matthes, Charles Petzold, Gerald Weinberg and Jack Ganssle all deepened my understanding. Jennifer Pahlka of Code for America, Tim O'Reilly, Sam Aaron of Sonic Pi and my talented Bay Area algorave buddy Thorsten Sideboard helped in different ways to expand my horizons. Tom Griffiths at Princeton, Neil Johnson at George Washington University and the modern Peripatetics Iain McGilchrist and George Dyson lent their infinitely extensible intelligences to helping me decode this flustered moment in time. The hospitality of Janet Siegmund's team in Magdeburg and especially the young scientists at Evelina Fedorenko's "EvLab" at MIT were inspirations, with special thanks due to Anna Ivanova at the latter. I'll be rooting for you all. Of course there are others too numerous to mention here by name, many of whom are in the book, many not, but all made significant contributions.

Closer to home, these Acknowledgments would be confined to some parallel universe full of unwritten books but for my priceless agent Emma Parry at Janklow & Nesbit, whose sage counsel, steadying hand and tireless energy have often kept me sane and approximately human when not much else could. I'd long wanted to work with Morgan Entrekin and his team at Grove, whose independence and integrity seem more precious now than they ever have, knowing my work would have a good home there. During production, Zoe Harris championed and put herself out for this more than anyone, at points working way beyond the call of duty and always providing a touchstone: there really isn't praise enough to confer on her.

Storytelling friends (though not all will love that description) who read or discussed parts of this book while I was trying to make it were, and always are, indispensable. My old *Face* colleague Chris Heath was the first to say, "You should write a book about that, I'd like to read it," while my great friend Simon Hattenstone (who makes an unnamed appearance in Chapter 17), John Best, Jacqueline Winspear and John Morell, Paul Willetts, Greg Takoudes and Dave Eggers all offered support and did me the honor of letting me talk it through with them at different stages. Whether they knew it or not, they were helping the material take shape. The fabulous and ever acute Bernard MacMahon read and offered comment on parts of the manuscript. Merope Mills,

Melissa Denes and Phil Olterman let me explore a few of the ideas in *The Guardian*, as did Sarah Baxter at the *Sunday Times* and Emma Duncan at *The Economist*. My children, Lotte L-S and Isaak Lewis-Smith, both writers, were and are a constant source of inspiration and feedback, always there as points of light if things get dark. Thanks to Emma Cline for the loan of an image: she knows which one.

This book took five years to make. It was the hardest thing I've ever done professionally, but also the most joyful and rewarding. There are two people I can't imagine it having been made without. One is my beloved wife Jan, who has that rare gift of making everyone around her better, including me, and whose patience and (often very real) sacrifice in standing unflinchingly with me while *Devil* was circling the runway, leaking fuel, there is simply no way to exaggerate. My only complaint against her is that she took so long to arrive in my life. But then, promptitude was never my strong suit, either. The other great gift was Nicholas Tollervey, who was with me nearly all the way through, offering advice, making introductions, adding depth to callow ideas and teaching by example, reading the manuscript at different stages and taking it upon himself to vet technical details and offer corrections where necessary (though, as before, any slips in that regard are mine). In the summer of 2023 he assisted me in delivering a keynote speech at the EuroPython conference in Prague, the merry squaring of a circle that felt like both an ending and a beginning. If all code was managed by people like Nicholas, the world would look very different indeed.

Andrew Smith
Brooklyn, NY

Select Bibliography

Abbate, Janet. *Recoding Gender: Women's Changing Participation in Computing*. Cambridge, MA: MIT Press, 2012.

Amos, Martyn. *Genesis Machines: The New Science of Biocomputing*. London: Atlantic Books, 2008.

Aspray, William. *John von Neumann and the Origins of Modern Computing*. Cambridge, MA: MIT Press, 1990.

Baron-Cohen, Simon. *The Essential Difference: Male and Female Brains and the Truth about Autism*. New York: Basic Books, 2003.

Baron-Cohen, Simon. *The Pattern Seekers: How Autism Drives Human Invention*. London: Allen Lane, 2020.

Bartik, Jean Jennings. *Pioneer Programmer: Jean Jennings Bartik and the Computer that Changed the World*. Edited by Jon T. Rickman and Kim D. Todd. Kirksville, MO: Truman State University Press, 2013.

Bartle, Richard. *Hearts, Clubs, Diamonds, Spades: Players Who Suit MUDs*. Colchester, UK: MUSE Ltd., 1996.

Bartle, Richard A. *Designing Virtual Worlds*. Syracuse, New York: New Riders Press, 2003.

Bartlett, Jamie. *The People vs. Tech: How the Internet Is Killing Democracy (and How We Can Save It)*. London: Ebury Press, 2018.

Bell, Ana. *Get Programming: Learn to Code with Python*. New York: Manning Publications, 2018.

Benavav, Aaron. *Automation and the Future of Work*. London: Verso, 2020.

Benjamin, Ruha. *Race after Technology*. Cambridge, UK: Polity Press, 2019.

Biggs, John. *Black Hat: Misfits, Criminals, and Scammers in the Internet Age*. Berkeley, CA: Apress, 2004.

Blakemore, Sarah-Jayne. *Inventing Ourselves: The Secret Life of the Teenage Brain*. New York: Public Affairs, 2018.

Boole, George. *The Continued Exercise of Reason: Public Addresses by George Boole*. Edited by Brendan Dooley. Cambridge, MA: MIT Press, 2018.

Boole, George. *An Investigation of the Laws of Thought*. Gearhart, OR: Watchmaker Publishing, 2010. First published in 1854.

Boole, George. *The Mathematical Analysis of Logic, Being an Essay Towards the Calculus of Deductive Reasoning*. London: George Bell, 1847.

Borges, Jorge Luis. *Labyrinths: Selected Stories & Other Writings*. New York: New Directions, 1962.

Bostrom, Nick. *Superintelligence: Paths, Dangers, Strategies*. Oxford: Oxford University Press, 2014.

Brandon, Richard. "The Problem in Perspective," in *Proceedings of the 1968 23rd Association for Computing Machinery (ACM) National Conference*. New York: ACM Press, 1968.

Bronson, Po. *The Nudist on the Late Shift: And Other True Tales of Silicon Valley*. New York: Random House, 1999.

Broussard, Meredith. *More Than a Glitch: Confronting Race, Gender and Ability Bias in Tech*. Cambridge, MA: MIT Press, 2023.

Calude, C. S., et al., *Unconventional Computing: 8th International Conference, UC 2009, Ponta Delgada, Portugal, September 7–11, 2009, Proceedings*. Berlin: Springer Verlag, 2009.

Ceder, Naomi. *The Quick Python Book* (3rd ed.). New York: Manning Publications, 2018.

Ceruzzi, Paul. "Electronics Technology and Computer Science, 1940–1975: A Coevolution." *Institute of Electrical and Electronics Engineers (IEEE) Annals of the History of Computing* 10, no.4. New York: IEEE, 1989.

Ceruzzi, Paul. *A New History of Modern Computing*. Cambridge, MA: MIT Press, 2021.

Chafkin, Max. *The Contrarian: Peter Thiel and Silicon Valley's Pursuit of Power*. New York: Penguin Press, 2021.

Chang, Emily. *Brotopia: Breaking Up the Boys' Club of Silicon Valley*. New York: Portfolio/Penguin, 2019.

Copeland, Jack, ed. *Colossus: The Secrets of Bletchley Park's Codebreaking Computers*. Oxford, UK: Oxford University Press, 2006.

Copeland, Jack. *Turing: Pioneer of the Information Age*. Oxford, UK: Oxford University Press, 2013.

Crowley Redding, Anna. *Google It: A History of Google*. New York: Feiwel and Friends, 2018.

Dijkstra, Edsger W. *Selected Writings on Computing: A Personal Perspective*. New York: Springer Verlag, 1982.

Dobson, Ann. *Touch Typing in Ten Hours*. Oxford: How To Books Ltd., 2002.

Dweck, Carol S. *Mindset: The New Psychology of Success*. New York: Ballantine Books, 2016.

Dyson, George. *Analogia: The Emergence of Technology beyond Programmable Control*. New York: Farrar, Straus & Giroux, 2020.

Dyson, George, *Darwin among the Machines: The Evolution of Global Intelligence*. New York: Basic Books, 1997.

Dyson, George. *Turing's Cathedral: The Origins of the Digital Universe*. New York: Vintage, 2012.

Ensmenger, Nathan. *The Computer Boys Take Over: Computers, Programmers, and the Politics of Technical Expertise*. Cambridge, MA: MIT Press, 2010.

Essinger, James. *Ada's Algorithm*. London: Melville House, 2017.

Eubanks, Virginia. *Automating Inequality: How High-Tech Tools Profile, Police, and Punish the Poor*. New York: St. Martin's Press, 2017.

Feldman Barrett, Lisa. *How Emotions Are Made: The Secret Life of the Brain*. Boston: Mariner Books, 2017.

Ferguson, Niall. *The Square and the Tower: Networks, Hierarchies and the Struggle for Global Power*. London: Allen Lane, 2017.

Flanagan, David. *JavaScript: Pocket Reference*. Sebastopol, CA: O'Reilly Media, 2012.

Flanagan, David. *JQuery: Pocket Reference*. Sebastopol, CA: O'Reilly Media, 2011.

Fleming, Ian. *Casino Royale*. New York: William Morrow, 2023.

Fleming, Ian. *From Russia with Love*. New York: William Morrow, 2023.

Fleming, Ian. *You Only Live Twice*. New York: William Morrow, 2023.

Fitzgerald, Michael. *Introducing Regular Expressions*. Sebastopol, CA: O'Reilly Media, 2012.

Foer, Franklin. *World without Mind: The Existential Threat of Big Tech*. New York: Penguin, 2018.

Foroohar, Rana. *Don't Be Evil: The Case against Big Tech*. London: Currency, 2019.

Forster, E. M. *The Machine Stops*. London: Penguin Modern Classics, 1928.

Friedman, Daniel P., and Felleisen, Matthias. *The Little Schemer*. Cambridge, MA: MIT Press, 1996.

Frommer, Franck. *How PowerPoint Makes You Stupid*. New York: The New Press, 2012.

Fry, Hannah. *Hello World: Being Human in the Digital Age*. New York: W. W. Norton, 2018.

Gadsby, Hannah. *Ten Steps to Nanette*. New York: Ballantine, 2022.

Galloway, Scott. *The Four: The Hidden DNA of Amazon, Apple, Facebook, and Google*. New York: Portfolio/Penguin, 2017.

Geraci, Robert M. *Apocalyptic AI: Visions of Heaven in Robotics, Artificial Intelligence and Virtual Reality*. Oxford, UK: Oxford University Press, 2010.

Gladwell, Malcolm. *Blink: The Power of Thinking without Thinking*. New York: Little, Brown and Company, 2005.

Gladwell, Malcolm. *Outliers: The Story of Success*. New York: Little, Brown and Company, 2008.

Goldstine, Herman. *The Computer from Pascal to von Neumann*. Princeton, NJ: Princeton University Press, 1972.

Griffiths, Tom, and Christian, Brian. *Algorithms to Live By: The Computer Science of Human Decisions*. New York: Picador, 2016.

Harris, Robert. *The Fear Index*. New York: Alfred A. Knopf, 2012.

Hertzfeld, Andy. *Revolution in the Valley: The Insanely Great Story of How the Mac Was Made*. Sebastopol, CA: O'Reilly Media, 2004.

Hicks, Marie. *Programmed Inequality: How Britain Discarded Women Technologists and Lost Its Edge in Computing*. Cambridge, MA: MIT Press, 2017.

Hillis, Daniel W. *The Pattern in the Stone: The Simple Ideas that Make Computers Work*. New York: Basic Books, 1998.

Hofstadter, Douglas R. *Gödel, Escher, Bach: An Eternal Golden Braid*. New York: Basic Books, 1979.

Isaacson, Walter. *The Innovators: How a Group of Hackers, Geniuses, and Geeks Created the Digital Revolution*. New York: Simon & Schuster, 2014.

Isaacson, Walter. *Steve Jobs*. New York: Simon & Schuster, 2011.

Ishiguro, Kazuo. *Klara and the Sun*. London: Faber & Faber, 2021.

Kidder, Tracy. *The Soul of a New Machine*. New York: Little, Brown and Company, 1981.

Kosseff, Jeff. *The Twenty-Six Words that Created the Internet*. Ithaca, NY: Cornell University Press, 2019.

Knuth, Donald E. *The Art of Computer Programming*. Volume 1: Fundamental Algorithms. Boston: Addison-Wesley Professional, 1997.

Knuth, Donald E. *Surreal Numbers*. Boston: Addison-Wesley Publishing, 1974.

Knuth, Donald E. *Things a Computer Scientist Rarely Talks About*. Stanford, CA: CSLI Publications, 2001.

Lanier, Jaron. *You Are Not a Gadget: A Manifesto*. New York: Vintage, 2011.

Lanzoni, Susan. *Empathy: A History*. New Haven, CT: Yale University Press, 2018.

Lasch, Christopher. *The Culture of Narcissism: American Life in an Age of Diminishing Expectations*. London: W. W. Norton, 1979.

Laster, Brett. *Professional Git*. New York: John Wiley & Sons, 2017.

Leveson, Nancy. *Engineering a Safer World: Systems Thinking Applied to Safety*. Cambridge, MA: MIT Press, 2016.

Levy, Stephen. *Hackers: Heroes of the Computer Revolution*. Sebastopol, CA: O'Reilly Media, 2010.

Light, Jennifer S. "When Computers Were Women." *Technology and Culture* 40, no. 3. Baltimore: Johns Hopkins University Press, 1999.

Lovelace, Augusta Ada, and L. F. Menabrea. *Sketch of the Analytical Engine Invented by Charles Babbage*. Quaternion Books, 2020.

MacCormick, John. *9 Algorithms that Changed the Future: The Ingenious Ideas that Drive Today's Computers*. Princeton, NJ: Princeton University Press, 2012.

MacHale, Desmond, and Cohen, Yvonne. *New Light on George Boole*. Cork, Ireland: Atrium, 2018.

Macrae, Norman. *John von Neumann: The Scientific Genius Who Pioneered the Modern Computer, Game Theory, Nuclear Deterrence, and Much More*. New York: Pantheon Press, 1992.

Malamud, Carl, and Pitroda, Sam. *Code Swaraj: Field Notes from the Standards Satyagraha*. Sebastopol, CA: Public.Resource.Org, Inc., 2018.

Margolis, Jane, and Fisher, Allan. *Unlocking the Clubhouse: Women in Computing*. Cambridge, MA: MIT Press, 2003.

Matthes, Eric. *Python Crash Course: A Hands-On, Project-Based Introduction to Programming*. San Francisco: No Starch Press, 2016.

McEwan, Ian. *Machines Like Me*. London: Jonathan Cape, 2019.

McGilchrist, Iain. *The Master and His Emissary: The Divided Brain and the Making of the Western World*. New Haven, CT: Yale University Press, 2009.

McGilchrist, Iain. *The Matter with Things: Our Brains, Our Delusions, and the Unmaking of the World*. London: Perspectiva Press, 2021.

McIlwain, Charlton D. *Black Software: The Internet and Racial Justice, from the AfroNet to Black Lives Matter*. Oxford, UK: Oxford University Press, 2020.

McLuhan, Marshall. *Understanding Me: Lectures and Interviews*. Cambridge, MA: MIT Press, 2003.

McNamee, Roger. *Zucked: Waking Up to the Facebook Catastrophe*. New York: Public Affairs, 2019.

Menn, Joseph. *The Cult of the Dead Cow: How the Original Hacking Supergroup Might Just Save the World*. New York: Public Affairs, 2019.

Menn, Joseph. *Fatal System Error: The Hunt for the New Crime Lords Who Are Bringing Down the Internet*. New York: Public Affairs, 2010.

Meyer, Eric A. *CSS: Pocket Reference*. Sebastopol, CA: O'Reilly Media, 2011.

Mickle, Tripp. *After Steve: How Apple Became a Trillion-Dollar Company and Lost Its Soul*. New York: William Morrow, 2022.

Minsky, Marvin. *The Emotion Machine: Commonsense Thinking, Artificial Intelligence and the Future of the Human*. New York: Simon & Schuster, 2006.

Misa, Thomas J., ed. *Gender Codes: Why Women Are Leaving Computing*. New York: John Wiley & Sons, 2010.

Mitchell, Ryan. *Web Scraping with Python: Collecting More Data from the Modern Web*. Sebastopol, CA: O'Reilly Media, 2018.

Mitnick, Kevin. *Ghost in the Wires: My Adventures as the World's Most Wanted Hacker*. New York: Little, Brown and Company, 2011.

Mlodinow, Leonard. *Emotional: How Feelings Shape Our Thinking*. New York: Pantheon, 2022.

Niederst Robbins, Jennifer. *HTML5: Pocket Reference*. Sebastopol, CA: O'Reilly Media, 2013.

Nixon, Robin. *Learning PHP, MySQL, JavaScript, CSS & HTML5: A Step-by-Step Guide to Creating Dynamic Websites*. Sebastopol, CA: O'Reilly Media, 2014.

Noble, Safiya Umoja. *Algorithms of Oppression: How Search Engines Reinforce Racism*. New York: NYU Press, 2018.

O'Keefe, David. *One Day in August: Ian Fleming, Enigma, and the Deadly Raid on Dieppe*. London: Icon Books, 2020.

O'Neil, Cathy. *Weapons of Math Destruction: How Big Data Increases Inequality and Threatens Democracy*. New York: Broadway Books, 2017.

Pasquale, Frank. *The Black Box Society: The Secret Algorithms that Control Money and Information*. Cambridge, MA: Harvard University Press, 2015.

Pein, Corey. *Live Work Work Work Die: A Journey into the Savage Heart of Silicon Valley*. New York, Metropolitan Books, 2017.

Petzold, Charles. *The Annotated Turing: A Guided Tour through Alan Turing's Historic Paper on Computability and the Turing Machine*. New York: John Wiley & Sons, 2008.

Petzold, Charles. *Code: The Hidden Language of Computer Hardware and Software*. Redmond, WA: Microsoft Press, 2000.

Pitt, Matthew, and Mauresmo, Kent. *WordPress Web Hosting: How to Use cPanel and Your Hosting Control Center*. Read2Learn.net, 2014.

Poldrack, Russell A. *The New Mind Reader: What Neuroimaging Can and Cannot Reveal about Our Thoughts*. Princeton, NJ: Princeton University Press, 2018.

Raymond, Eric S. *The Cathedral & the Bazaar*. Sebastopol, CA: O'Reilly Media, 2001.

Raynor, Tim. *Hacker Culture and the New Rules of Innovation*. Oxford: Routledge, 2018.

Rosenberg, Scott. *Dreaming in Code*. New York: Three Rivers Press, 2007.

Rothblatt, Martine. *Virtually Human: The Promise and the Peril of Digital Immortality*. New York: St. Martin's Press, 2014.

Saunders, George. *Congratulations, by the Way: Some Thoughts on Kindness*. New York: Random House, 2014.

Sax, David. *The Revenge of Analog: Real Things and Why They Matter*. New York: Public Affairs, 2016.

Seife, Charles. *Zero: The Biography of a Dangerous Idea*. London: Penguin Books, 2000.

Shasha, Dennis, and Lazere, Cathy. *Out of Their Minds: The Lives and Discoveries of 15 Great Computer Scientists*. New York: Copernicus, 1995.

Siegel, Dan. *Mind: A Journey to the Heart of Being Human*. New York: W. W. Norton & Company, 2016.

Silberman, Steve. *Neurotribes: The Legacy of Autism and How to Think Smarter about People Who Think Differently*. London: Allen & Unwin, 2015.

Smith, Andrew. *Moondust: In Search of the Men Who Fell to Earth*. London: Bloomsbury, 2005.

Smith, Andrew. *Totally Wired: The Rise and Fall of Josh Harris and the Great Dotcom Swindle*. New York: Grove Atlantic, 2019.

Song, Jimmy. *Programming Bitcoin: Learn How to Program Bitcoin from Scratch*. Sebastopol, CA: O'Reilly Media, 2019.

Stephens-Davidowitz, Seth. *Everybody Lies: Big Data, New Data, and What the Internet Can Tell Us about Who We Really Are*. New York: HarperCollins, 2017.

Stoll, Cliff. *The Cuckoo's Egg: Tracking a Spy through the Maze of Computer Espionage*. New York: Pocket Books, 1981.

Stubbs, David. *Mars by 1980: The Story of Electronic Music*. London: Faber & Faber, 2018.

Sweigart, Al. *Automate the Boring Stuff with Python: Practical Programming for Total Beginners*. San Francisco: No Starch Press, 2015.

Taylor, Astra. *Taking Back Power and Culture in the Digital Age*. London: Fourth Estate, 2014.

Thomas, Douglas. *Hacker Culture*. Minneapolis: University of Minnesota Press, 2002.

Thompson, Clive. *Smarter Than You Think: How Technology Is Changing Our Minds for the Better*. New York: Penguin Books, 2013.

Thompson, Helen. *Unthinkable: An Extraordinary Journey through the World's Strangest Brains*. London: John Murray, 2018.

Thorpe, D. R. *Supermac: The Life of Harold Macmillan*. London: Chatto & Windus, 2010.

Tollervey, Nicholas. *Programming with MicroPython*. Sebastopol, CA: O'Reilly Media, 2018.

Tollervey, Nicholas. *Python in Education: Teach, Learn, Program*. Sebastopol, CA: O'Reilly Media, 2015.

Turing, Alan M. *The Essential Turing: Seminal Writings in Computing, Logic, Philosophy, Artificial Intelligence, and Artificial Life plus The Secrets of Enigma*. Edited by B. Jack Copeland. Oxford: Clarendon Press, 2004.

Turkle, Sherry. *The Second Self: Computers and the Human Spirit*. Cambridge, MA: MIT Press, 1984.

Ullman, Ellen. *The Bug*. London: Pushkin Press, 2003.

Ullman, Ellen. *Close to the Machine: Technophilia and Its Discontents*. London: Pushkin Press, 1997.

Ullman, Ellen. *Life in Code: A Personal History of Technology*. New York: Farrar, Straus & Giroux, 2017.

Vaughan, Lee. *Impractical Python Projects: Playful Programming Activities to Make You Smarter*. San Francisco: No Starch Press, 2019.

Vonnegut, Kurt. *Player Piano*. New York: Scribner, 1952.

von Neumann, John, and Kurzweil, Ray. *The Computer and the Brain*. New Haven, CT: Yale University Press, 2012.

Wang, Wallace. *Beginning Programming for Dummies*. Hoboken, NJ: John Wiley & Sons, 2007.

Weinberg, Gerald. *The Psychology of Computer Programming*. New York: Dorset House, 1998.

Weizenbaum, Joseph. *Computer Power and Human Reason: From Judgement to Calculation*. San Francisco: W. H. Freeman and Company, 1976.

Weizenbaum, Joseph. *Islands in the Cyberstream*. Sacramento, CA: Litwin Books, 2015.

Wiener, Anna. *Uncanny Valley: A Memoir*. New York: Farrar, Straus and Giroux, 2020.

Wilson, Scott, Cottle, David, and Collins, Nick. eds., *The SuperCollider Book*. Cambridge, MA: The MIT Press, 2011.

Winterson, Jeanette. *12 Bytes*. New York: Grove Atlantic, 2021.

Wittgenstein, Ludwig. *Culture and Value*. Chicago: University of Chicago Press, 1977.

Wittgenstein, Ludwig. *Philosophical Investigations* (3rd ed.). London: Basil Blackwell & Mott, Ltd., 1958.

Wroblewski, Luke. *Site-Seeing: A Visual Approach to Web Usability*. New York: Hungry Minds, 2002.

Zak, Paul J. *The Moral Molecule: How Trust Works*. New York: Plume Books, 2013.

Index